Ross and Wilson

ANATOMY and PHYSIOLOGY

colouring and workbook

Commissioning Editor: Ninette Premdas
Development Editor: Clive Hewat
Project Manager: Elouise Ball
Designer: Kirsteen Wright
Illustration Manager: Merlyn Harvey

Ross and Wilson
ANATOMY and PHYSIOLOGY
colouring and workbook | 3rd EDITION

Anne Waugh BSc(Hons) MSc CertEd SRN RNT FHEA
Senior Lecturer and Senior Teaching Fellow,
School of Nursing, Midwifery and Social Care,
Edinburgh Napier University, Edinburgh, UK

Allison Grant BSc PhD RGN
Lecturer, School of Biological and Biomedical Sciences,
Glasgow Caledonian University, Glasgow, UK

Illustrations by **Graeme Chambers**

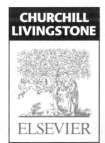

CHURCHILL LIVINGSTONE

ELSEVIER

Edinburgh London New York Oxford Philadelphia St Louis Sydney Toronto 2010

CHURCHILL
LIVINGSTONE
ELSEVIER

© 2004, Elsevier Limited. All rights reserved
© 2006 Elsevier Limited. All rights reserved
Third edition © 2010, Elsevier Limited. All rights reserved.

ISBN - 9780702032264
  Reprinted 2011 (twice), 2012

British Library Cataloguing in Publication Data
A catalogue record for this book is available from the British Library

Library of Congress Cataloging in Publication Data
A catalog record for this book is available from the Library of Congress

Notice

Knowledge and best practice in this field are constantly changing. As new research and experience broaden our knowledge, changes in practice, treatment and drug therapy may become necessary or appropriate. Readers are advised to check the most current information provided (i) on procedures featured or (ii) by the manufacturer of each product to be administered, to verify the recommended dose or formula, the method and duration of administration, and contraindications. It is the responsibility of the practitioner, relying on their own experience and knowledge of the patient, to make diagnoses, to determine dosages and the best treatment for each individual patient, and to take all appropriate safety precautions. To the fullest extent of the law, neither the Publisher nor the Authors assumes any liability for any injury and/or damage to persons or property arising out of or related to any use of the material contained in this book.

The Publisher

Printed in China

Contents

Preface

Ross and Wilson has been a core text for students of anatomy and physiology for over 40 years. Although this companion text has been extensively revised to match the eleventh edition of the main text, providing varied learning activities to facilitate and reinforce learning, it can also be used to support any general anatomy and physiology course.

The systems approach used in the main text forms the framework for the exercises, many of which are based on clear illustrations of body structure and functions. A variety of exercises have been devised to maintain interest and provide choice. The section on 'How to use this book', p. ix, explains how the icons and exercises are used in the text.

We hope that you will find this book a stimulating and useful companion to your anatomy and physiology studies, including those times when revision is required. We are always delighted to receive feedback, especially from students, so please continue to send your comments to us via the publishers.

We would have been unable to prepare the new edition without the help and support of many others, including Graeme Chambers who has, as always, patiently revised and created new artwork for this edition. Several people at Churchill Livingstone have also provided encouragement and support in preparation of the new edition and, in particular, we would like to thank Ninette Premdas and Clive Hewat.

We would also like to thank our families, Andy, Michael, Seona and Struan for their continuing help and support with this venture.

Edinburgh, 2009

Anne Waugh
Allison Grant

How to use this book
Icons and exercises

 Colouring: identify and colour structures on diagrams.

 Labelling: identify and label structures on diagrams.

 Matching: match statements with reasons; structures with functions; key choices with blanks in a paragraph; and organs on diagrams.

Combinations of these activities are also used to provide variety in the text.

 Multiple-choice questions: identify the correct option from a list of four. Where there is more than one correct option, this is indicated in the question.

 Completion: identify the missing word to complete paragraphs explaining body structure and functions.

 Definitions: explain the meaning of a common anatomical or physiological term.

 Pot luck: a variety of other exercises is also used to facilitate learning. Simple guidance about completion is provided.

The body as a whole

The human body is complex, like a highly technical and sophisticated machine. Although it operates as a single entity, it is made up of several parts that work interdependently. This chapter will help you learn about the major systems and control mechanisms that maintain integrated body functioning.

LEVELS OF STRUCTURAL COMPLEXITY

 Matching

1. Match the key choices below with the labels on Figure 1.1.

Key choices:
System level
Cellular level
Organ level
The human being
Chemical level
Tissue level

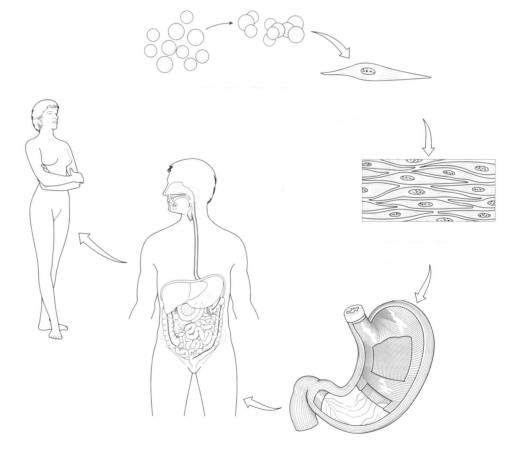

Figure 1.1 The levels of structural complexity

Matching

2. Using the list of key choices on the previous page, complete Table 1.1.

Level of structural complexity	Characteristics
	Comprises many systems that work interdependently to maintain health
	Carry out a specific function and are composed of different types of tissue
	The smallest independent units of living matter
	Consist of one or more organs and contribute to one or more survival needs of the body
	Atoms and molecules that form the building blocks of larger substances
	A group of cells with similar structures and functions

Table 1.1 Levels of structural complexity and their characteristics

 MCQs

3. The study of body structure and the physical relationships between body parts is: _____.

   a. Anatomy　　　　**b.** Physiology　　　　**c.** Pathology　　　　**d.** Pathophysiology.

4. The study of abnormalities and how they affect body function is: _____.

   a. Anatomy　　　　**b.** Physiology　　　　**c.** Pathology　　　　**d.** Pathophysiology.

5. Which list correctly describes the order of chemical complexity starting with the most simple? _____.

   a. Molecules, atoms, organs, tissues　　　**c.** Atoms, molecules, tissues, organs
   b. Tissues, organs, molecules, atoms　　　**d.** Organs, tissues, molecules, atoms.

6. The smallest independent units of living matter in the body are: _____.

   a. Tissues　　　　**b.** Complex molecules　　　　**c.** Organs　　　　**d.** Cells.

THE INTERNAL ENVIRONMENT AND HOMEOSTASIS

 Completion

7. Complete the paragraph correctly by crossing out the wrong statements.

The **internal/external** environment surrounds the body and provides oxygen and nutrients required by the body cells. The **cell membrane/skin** provides a barrier between the **solid/dry** external environment and the **watery/gaseous** internal environment. The **internal/external** environment is the medium in which all body cells exist. Cells are bathed in fluid called **interstitial/intracellular** fluid, also known as **lymph/tissue fluid**. The **cell membrane/tissues** provide(s) a potential barrier to substances entering or leaving the cell. **It/they** prevent(s) **small/large** molecules moving between the cell and interstitial fluid. **Small/large** molecules can usually pass through the membrane and therefore the chemical composition of the fluid inside the cell is **the same/different** from that outside. This property is known as **osmosis/selective permeability**.

Completion

8. Identify the three components of the negative feedback mechanism in the central heating system shown in Figure 1.2:

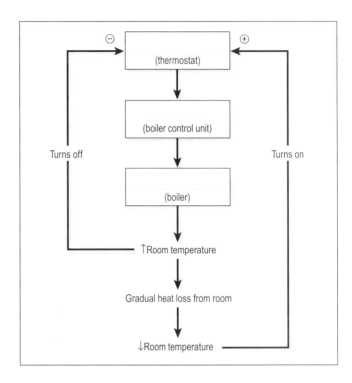

Figure 1.2 Example of a negative feedback mechanism: control of room temperature by a domestic boiler

9. Fill in the blanks in the paragraph below to describe how a negative feedback mechanism operates using body temperature as an example.

The composition of the internal environment is maintained within narrow limits, and this fairly constant state is

called _____. In systems controlled by negative feedback mechanisms, the effector response _____

the effect of the original stimulus. When body temperature falls below the preset level, specialized temperature-

sensitive nerve endings act as _____ and relay this information to cells in the hypothalamus of the brain

that form the _____. This results in activation of _____ responses that raise body

temperature. When body temperature returns to the _____ range again, the temperature-sensitive nerve

endings no longer stimulate the cells in the hypothalamus and the heat conserving mechanisms are switched off.

Pot luck

10. List two physiological responses that will counteract a fall in body temperature:

- _____

- _____.

Pot luck

11. Identify two physiological responses that occur in response to a rise in body temperature.

    • _____   • _____.

12. State four other physiological variables that are controlled by negative feedback:

    • _____   • _____

    • _____   • _____.

13. Briefly outline how a positive feedback mechanism operates.

    _____

    _____.

SURVIVAL NEEDS OF THE BODY

Pot luck

14. Which system is concerned with:

    a. Intake of oxygen? _____

    b. Intake of nutrients? _____

    c. Protection against the external environment? _____.

15. Which system excretes each of the following waste products?

    a. Faeces: _____

    b. Urine: _____

    c. Carbon dioxide: _____.

16. Briefly outline the difference between specific and non-specific defence mechanisms.

    _____

    _____

    _____

    _____.

17. True or false? Circle the correct answer for each statement.

    a. Most body movement is not under conscious control   **(T/F)**

    b. Skeletal muscles move the joints   **(T/F)**

    c. Skeletal muscles are attached to bones by tendons   **(T/F)**

    d. Red blood cells are also known as leukocytes   **(T/F)**

    e. The smallest blood vessels are capillaries and have very thin walls   **(T/F)**.

Completion

18. Complete the paragraph below describing the function of the female reproductive system.

The childbearing years begin at _____ and end at the _____. During this time an _____

matures in the ovary about every _____ days. If _____ takes place it embeds itself in the _____

and grows to maturity during pregnancy, or _____, in about _____ weeks. If fertilization does not occur

it is shed with the uterine lining, accompanied by bleeding, called _____.

Definitions

Define the following terms:

19. Afferent _____.

20. Efferent _____.

21. Antigen _____.

22. Allergic reaction _____.

Colouring, matching and labelling

23. Colour and match the following structures on Figure 1.3:

○ Heart
○ Blood vessels

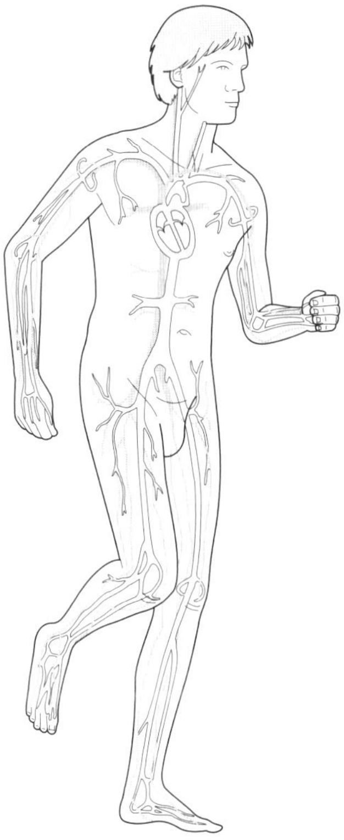

Pot luck

24. The fluid part of the blood is known as _____.

25. The blood volume in adults is approximately _____.

26. Blood vessels that carry blood away from the heart are _____.

27. The normal pulse rate is around _____ beats per minute.

Figure 1.3 The circulatory system

Colouring and labelling

28. Colour and label the following parts of the lymphatic system on Figure 1.4:

> Lymph nodes
> Lymph vessels

29. Label the heart on Figure 1.4.

30. What is the function of lymph nodes?

_____.

31. Which white blood cells involved in immunity mature

in the lymphatic system? _____.

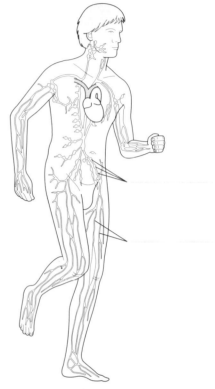

Figure 1.4 The lymphatic system: lymph nodes and vessels

  Colouring, matching and labelling

32. Colour and match the following parts of the nervous system shown on Figure 1.5:

> ○ Central nervous system
> ○ Peripheral nervous system

33. Label the structures indicated on Figure 1.5.

34. The very fast withdrawal of a finger from a very hot

surface is an example of a _____.

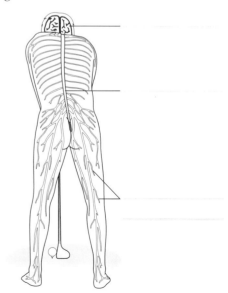

Figure 1.5 The nervous system

Completion

35. Fill in the blanks in the paragraph below to provide an overview of the endocrine system.

The endocrine system consists of a number of _____ in various parts of the body. The glands synthesize

and secrete chemical messengers called _____ into the _____. These chemicals stimulate

_____. Changes in hormone levels are usually controlled by _____

mechanisms. The endocrine system, in conjunction with part of the _____ system, controls

_____ body function. Changes involving the latter system are usually _____ while those of

the endocrine system tend to be _____ and precise.

36. Complete Table 1.2 by inserting the appropriate sensory organs for each of the special senses.

Special sense	Related sensory organ
Sight	
Hearing	
Balance	
Smell	
Taste	

Table 1.2 The special senses and their related sensory organs

Matching and labelling

37. Match the structures listed below with the labels on Figure 1.6:

> Bronchus
> Lung
> Trachea
> Larynx
> Nasal cavity
> Pharynx
> Oral cavity

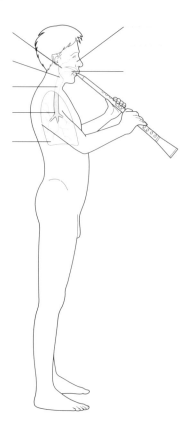

Figure 1.6   The respiratory system

  Colouring and labelling

38. Colour and label the organs of the digestive system shown on Figure 1.7.

39. Circle the accessory organs associated with the digestive system.

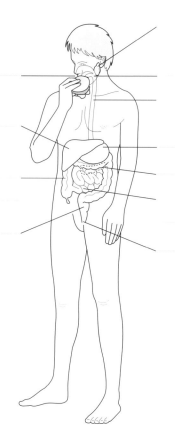

Figure 1.7   The digestive system

40. Name the chemical substances that digest food. _____ .

41. Colour and label the organs of the urinary system on Figure 1.8:

| Bladder |
| Kidney |
| Urethra |
| Ureter |

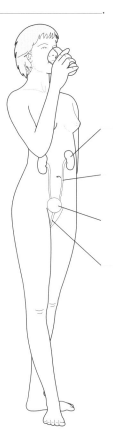

Figure 1.8 The urinary system

42. In which organ is urine formed? _____ .

43. Name the chemical substances that control water balance. _____ .

 Colouring

44. Colour the skeletal muscles on Figure 1.9.

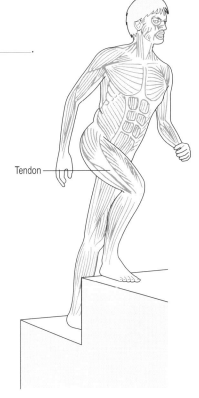

Tendon

Figure 1.9 The skeletal muscles

Colouring and matching

45. Colour and match the structures listed below with those identified on
    Figure 1.10:

- ○ Vagina
- ○ Ovary
- ○ Uterine tube
- ○ Uterus
- ○ Testis
- ○ Prostate gland
- ○ Penis
- ○ Deferent duct

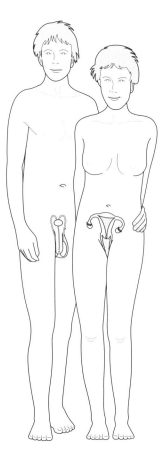

Figure 1.10   The reproductive systems: male and female

 Definitions

Define the following terms:

46. Anabolism _____.

47. Catabolism _____.

48. Micturition _____.

49. Defaecation _____.

Matching

50. Match the items in list A with the definitions in list B.

List A

Acute	Sign
Acquired	Symptom
Chronic	Syndrome
Congenital	

List B

a. An abnormality described by the patient: _____

b. A disorder with which one is born: _____

c. A disorder which develops after birth: _____

d. A long-standing disorder which cannot be cured: _____

e. A collection of signs and symptoms which usually occur together: _____

f. A disease with sudden onset: _____

g. An abnormality seen or measured by people other than the patient: _____ .

Definitions

Define the following terms:

51. Aetiology _____

_____ .

52. Pathogenesis _____

_____ .

53. Prognosis _____

_____ .

54. Idiopathic _____

_____ .

Electrolytes and body fluids

To understand how the tissues, organs and systems of the body work, their fundamental building blocks must be studied. This chapter covers basic chemistry and the structures and functions of important biological molecules.

ATOMS, MOLECULES AND COMPOUNDS

 Definitions

Define the following terms:

1. Atom _____

   _____.

2. Compound _____

   _____.

3. Element _____

   _____.

 Labelling and completion

4. Figure 2.1 shows the basic structure of an atom.
   Label the structures shown.

5. Fill in the blank circles to show the maximum number of electrons in each energy level.

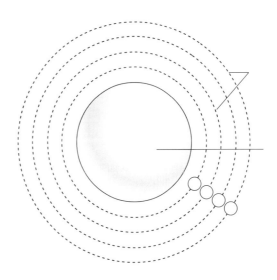

Figure 2.1 The atom

Matching

6. Match each of the statements in the box with the appropriate subatomic particle in list A below.

List A

Electron.....................
Proton.......................
Neutron....................

a. positively charged _____

b. found in the atomic nucleus _____

c. carries one mass unit _____

d. orbits atomic nucleus _____

e. negatively charged _____

f. electrically neutral _____

g. the number of these particles in the atom is called the atomic number _____

h. the sum of these two particles in the atom is called the atomic weight _____

i. is responsible for forming bonds between atoms

_____.

Completion

7. Complete the grid in Figure 2.2 by filling in the atomic numbers and atomic weights of the atoms shown.

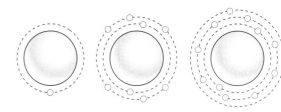

Figure 2.2 The atomic structures of hydrogen, oxygen and sodium

	Hydrogen	Oxygen	Sodium
Atomic number			
Atomic weight			

? MCQs

8. Isotopes are atoms of the same element with different numbers of: _____.

   a. Protons and electrons   b. Electrons   c. Neutrons   d. Protons and neutrons.

9. Two isotopes of the same element will differ in atomic: _____.

   a. Charge   b. Weight   c. Number   d. Energy.

10. Tritium is an isotope of: _____.

    a. oxygen   b. sodium   c. chlorine   d. hydrogen.

11. An organic substance, by definition, contains: _____.

    a. water   b. oxygen   c. carbon   d. iron.

Completion

12. The following paragraph describes chemical bonds. Complete it by crossing out the incorrect options in bold.

The commonest type of atomic bond is called a(n) **ionic/hydrogen/covalent** bond. This is a **stable/unstable** bond, joining atoms **firmly/loosely** together, and the molecules formed are **electrically charged/electrically neutral**. This bond is formed when **electrons/neutrons/protons** are **shared between atoms/donated from one atom to another**. An example of a molecule with such a bond is **water/sodium chloride/sodium bicarbonate**.

The next most common form of atomic bond is the **ionic/hydrogen/covalent** bond. This is **more/less** stable than the bond identified above. It is formed when **electrons/neutrons/protons** are **shared between atoms/donated from one atom to another**. When molecules containing this type of bond are dissolved in water, the bonds break to release **electrons/atoms/ions**. Such a substance is called a(n) **electrolyte/buffer/acid**.

Pot luck

13. List three functions of electrolytes:

- _____

- _____

- _____ .

MCQs

14. A solution containing more hydrogen ions than hydroxyl ions is: _____ .

    a. a buffer   **b.** pH neutral   **c.** acid   **d.** inorganic.

15. An alkaline solution is characterized by high levels of which ion? _____ .

    a. Bicarbonate   **b.** Hydroxyl   **c.** Hydrogen   **d.** Sodium.

16. Which of the following is true? _____ .

    a. An acid solution has a higher pH than an alkaline solution
    b. A strongly acidic solution has a higher pH than a weaker one
    c. There are no ions in a neutral solution
    d. An alkaline solution has a higher pH than an acid one.

17. What is the function of an acid buffer in the body? _____ .

    a. It mops up hydroxyl ions and increases pH   **c.** It mops up hydroxyl ions and decreases pH
    b. It mops up hydrogen ions and decreases pH   **d.** It mops up hydrogen ions and increases pH.

? Pot luck

18. Which two organs are most important in maintaining the acid–base balance in the body by adjusting excretion of excess acid or base?

    • _____

    • _____.

19. Define the term alkalosis:

    _____

    _____.

20. Write down the equation that represents the conversion of carbon dioxide to bicarbonate in body fluids.

    _____

    _____.

↰ Matching

21. Each of the following situations listed in the box, if uncompensated, will tend to tip the body into acidosis or alkalosis. Match each situation to acidosis or alkalosis.

    | Acidosis...................
    | Alkalosis..................

    a. reduced renal excretion of bicarbonate _____

    b. reduced ventilation _____

    c. exhaustion of bicarbonate _____

    d. hyperventilation _____

    e. reduced renal excretion of hydrogen ions _____

    f. prolonged vomiting _____.

IMPORTANT BIOLOGICAL MOLECULES

 Completion

22. For each of the statements in the left hand column of Table 2.1, decide to which of the classes of biological molecules it belongs (there may be more than one) and tick the appropriate boxes in the table.

	Carbohydrates	Proteins	Nucleotides	Lipids
Building blocks are amino acids				
Contain carbon				
Molecules joined with glycosidic linkages				
Used to build genetic material				
Building blocks are monosaccharides				
Contain glycerol				
Contain hydrogen				
Molecules joined together with peptide bonds				
Strongly hydrophobic				
Built from sugar unit, phosphate group and base				
Enzymes are made from these				
Contain oxygen				

Table 2.1 Characteristics of some important biological molecules

23. Complete the chemical structure in Figure 2.3, which represents a molecule of glucose, by identifying the atoms at the points shown.

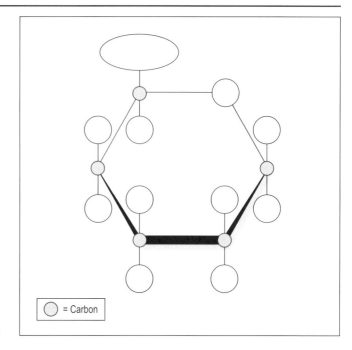

( ) = Carbon

Figure 2.3   Structure of a glucose molecule

 Pot luck

24. List the four main functions of the carbohydrates.

- _____

- _____

- _____

- _____ .

Labelling

25. Figure 2.4 shows the general structure of amino acids, the building block of proteins. Complete it by inserting the appropriate chemical symbols in the spaces provided.

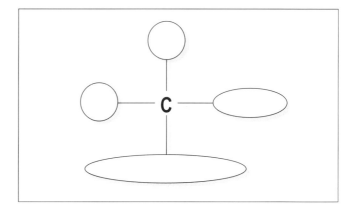

Figure 2.4   Amino acid structure

Pot luck

26. Proteins are used in the body in many ways. Identify which of the following are composed (at least primarily) of protein by circling the item(s).

Insulin
Haemoglobin
The cell membrane
Glycogen
Vitamin K
Adenosine triphosphate
Deoxyribonucleic acid
Adipose tissue
Antibodies
Enzymes
Sucrose
Collagen

 Completion

27. Complete Figure 2.5, which shows the general structure of a fat, by labelling the molecular groups shown.

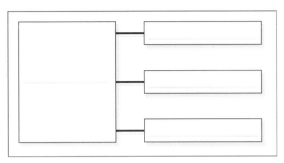

Figure 2.5   General structure of a fat molecule

 Pot luck

28. The following paragraph relates to lipids, but contains *eight* errors. Find the errors and correct them.

Lipids are a group of substances that are all strongly hydrophobic, meaning they dissolve well in water. Although the different types of lipid are chemically diverse, they all contain carbon, hydrogen, oxygen and sodium atoms. There are different types of lipid. One type is the fat, stored in the body's adipose tissues as a source of energy. Fat is a more efficient source of energy than carbohydrate, as it releases more energy when broken down. In addition, fat stored under the skin insulates the body and protects underlying structures. A fat molecule consists of two fatty acid molecules linked to a molecule of phospholipid. Other types of lipids include certain vitamins, e.g. vitamin C, an example of a fat-soluble vitamin. Some hormones are lipids, including the steroid hormones. Other lipids include the phospholipids, which form an integral part of the cell's DNA.

Labelling

29. Figure 2.6 shows the structure of ATP and its interconversion with ADP. Label the structures indicated.

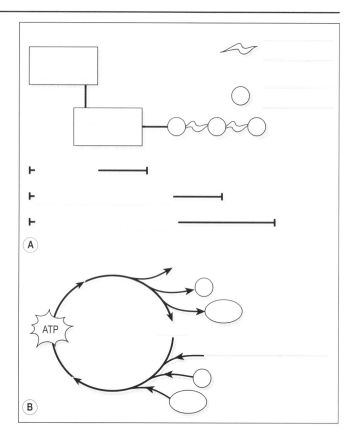

Figure 2.6 ATP and ADP. A. Structures. B. Conversion cycle

 Pot luck

30. The following statements concern the nature and function of enzymes. One is correct and three are false. Identify the false statements and write a correct version below.

   a. Enzymes are proteins used in the body to regulate (speed up or slow down) chemical reactions in body cells.

   _____

   _____

   b. Enzymes are not normally changed by the reactions they participate in, meaning that an enzyme molecule can catalyse the same reaction over and over again.

   _____

   _____

   c. Enzymes can catalyse the production of larger molecules from smaller ones, and this is called synthesis or catabolism.

   _____

   _____

   d. Enzymes are multi-functional, in that the body uses only a very few enzymes to regulate its multitude of chemical reactions. Each enzyme therefore is capable of catalysing many different types of reaction, involving different types of reactants.

   _____

   _____

MOVEMENT OF SUBSTANCES WITHIN THE BODY

 MCQs

31. Which of the following is true regarding a substance moving down its concentration gradient? _____.

    a. It requires energy

    b. Diffusion always involves such movement

    c. Substances cannot move down a concentration gradient

    d. It cannot occur across a barrier such as a cell membrane.

32. Which of the following physiological processes involves diffusion? _____.

    a. Gas exchange in the alveoli

    b. Exchange of sodium and potassium ions across cell membranes

    c. Water movement in and out of cells

    d. Movement of molecules that requires a supply of ATP.

33. Movement of water molecules occurs by: _____.

    a. Diffusion   **b.** Active transport   **c.** Dilution   **d.** Osmosis.

34. Which of the following statements regarding water movement is true (choose all that apply)? _____.

    a. A hypertonic solution contains more solute (dissolved particles) than a hypotonic one

    b. Water moves from a hypotonic solution into a hypertonic solution (assuming no barrier to water movement exists)

    c. Red blood cells placed in a beaker of pure water will shrink because water will leave the cells across the semipermeable cell membrane

    d. If two solutions on either side of a semipermeable membrane are isotonic to each other, it means no net movement of water molecules will occur.

BODY FLUIDS

 Pot luck

35. What percentage of body mass in an average adult is water? _____.

36. Which of the following is associated primarily with the intracellular environment (circle all that apply)?

Sodium	Urine
Cytoplasm	Glomerular filtrate
Cerebrospinal fluid	Lymph
Synovial fluid	Potassium
Plasma	Blood
Saliva	ATP

The cells, tissues and organization of the body

Cells are the smallest functional units of the body. Groups of similar cells form tissues, each of which has a distinct and specialized function. This chapter will help you learn about the structure of cells and characteristics of different types of tissue. The last section considers the organization of the body including anatomical terminology, the skeleton and the body cavities.

THE CELL: STRUCTURE AND FUNCTIONS

  Colouring and labelling

1. Colour and label the intracellular organelles identified on Figure 3.1.

2. Label the plasma membrane on Figure 3.1.

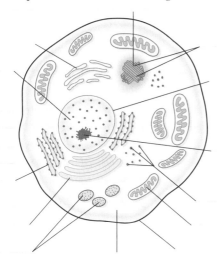

Figure 3.1 The simple cell

 Matching

3. Match the organelles from the list of key choices below with their functions in Table 3.1:

Key choices:	
Lysosomes	Microfilaments
Ribosomes	Microtubules
Nucleus	Golgi apparatus
Smooth endoplasmic reticulum	Mitochondria
Rough endoplasmic reticulum	

Organelle	Function
	The largest organelle, directs the activities of the cell
	Sausage-shaped structures, often described as the powerhouse of the cell. Sites of aerobic respiration
	Tiny granules consisting of RNA and protein that synthesize proteins for use within cells
	Proteins exported from cells are manufactured here
	Lipids and steroid hormones are synthesized here
	Stacks of closely flattened membranous sacs that form membrane-bound granules called secretory vesicles
	Secretory vesicles that contain enzymes for the breakdown of large cellular wastes, e.g. fragments of old organelles
	The tiny strands of protein that provide the structural support and shape of a cell
	Contractile proteins involved in movement of cells and of organelles within cells

Table 3.1 Intracellular organelles and their functions

Pot luck

4. There are seven errors in the paragraph below describing the structure of cell membranes. Find the errors and correct them.

The plasma membrane consists of two layers of phospholipids with some carbohydrate molecules embedded in them. The phospholipid molecules have a head which is electrically charged and hydrophilic (meaning water hating) and a tail that has no charge and is hydrophobic. The hormone cholesterol is also present. The phospholipid bilayer is arranged like a sandwich with the hydrophilic heads on the inside and the hydrophobic tails on the outside. These differences influence the passage of substances across the membrane. In motile cells, bodies project from the plasma membrane and include long cilia which permit movement of the cell.

  Labelling and colouring

5. Label the stages of mitosis by completing the shaded boxes on Figure 3.2.

6. Colour the genetic material shown on the parts of Figure 3.2.

7. Label the cellular structures seen on light microscopy during mitosis.

 Completion

8. Complete the paragraph about the cell cycle by crossing out the incorrect options.

Most body cells have **23/46** chromosomes and divide by **mitosis/meiosis**. The daughter cells of mitosis are genetically **different/identical**. Formation of gametes takes place by **mitosis/meiosis** and the daughter cells are genetically **different/identical**. The period between two cell divisions is known as the **cell/chromosome** cycle, which has two stages, the M phase and interphase. **The M phase/interphase** is the longer stage. Interphase has **three/four** separate stages. The stage where there is most cell growth is the **first gap phase/second gap phase**. The stage at which chromosomes replicate is the **second gap phase/S phase**. Mitosis has **three/four** identifiable stages.

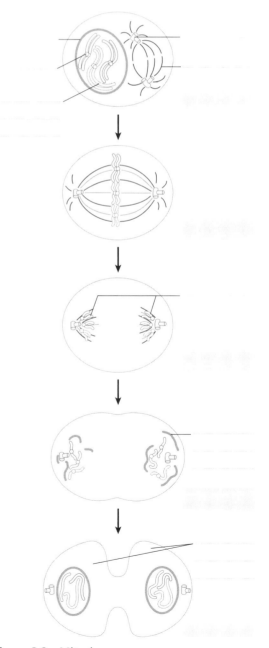

Figure 3.2 Mitosis

Definitions

Define the following terms:

9. Active transport _____

_____ .

10. Passive transport _____

_____ .

? MCQs

11. Which forms of transport across cell membranes require energy? (choose all that apply) _____ .

    a. Active transport   **b.** Facilitated diffusion   **c.** Osmosis   **d.** Bulk transport.

12. The number of carrier molecules determines the rate of: _____ .

    a. Active transport   **b.** Facilitated diffusion   **c.** Osmosis   **d.** Bulk transport.

13. Movement of water across a cell membrane down its concentration gradient refers to: _____ .

    a. Active transport   **b.** Facilitated diffusion   **c.** Osmosis   **d.** Bulk transport.

14. Movement of a small molecule up its concentration gradient occurs during: _____ .

    a. Active transport   **b.** Diffusion   **c.** Osmosis   **d.** Bulk transport.

15. Which forms of movement only take place across membranes? (choose all that apply) _____ .

    a. Active transport   **b.** Diffusion   **c.** Osmosis   **d.** Bulk transport.

16. Gas molecules move across cell membranes by: _____ .

    a. Active transport   **b.** Diffusion   **c.** Osmosis   **d.** Bulk transport.

↰ Matching

17. Match the key choices to the spaces in the paragraph below to describe bulk transport.

Key choices:
Exocytosis
Digest
Pinocytosis
Vacuole
Phagocytosis
Plasma membrane
Enzymes
Lysosomes

Transfer of large particles across the plasma membrane into the cell occurs by

_____ and _____ . The particles are engulfed by extensions of the

_____ that enclose them forming a membrane-bound

_____ . Then _____ adhere to the cell membrane releasing

_____ that _____ the contents. Extrusion of waste materials by

the reverse process is called _____ .

TISSUES

  Colouring, matching and labelling

18. Name the types of epithelial tissues shown in Figure 3.3.

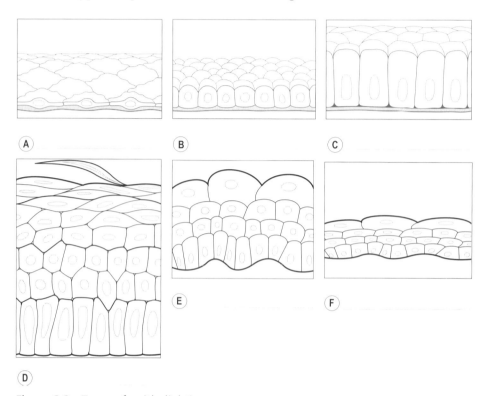

Figure 3.3 Types of epithelial tissue

19. On each type of epithelial tissue shown in Figure 3.3, colour and match where shown:

○ Nuclei
○ Epithelial cells
○ Basement membranes

20. Identify one body organ where the tissue on Figure 3.3E is found. _____.

21. Outline the function of the tissue shown in Figure 3.3E.

_____.

  Colouring, matching and labelling

22. Name each type of connective tissue shown in Figure 3.4.

23. Colour the matrix on each part of Figure 3.4.

24. Label the cells and fibres on each part of Figure 3.4.

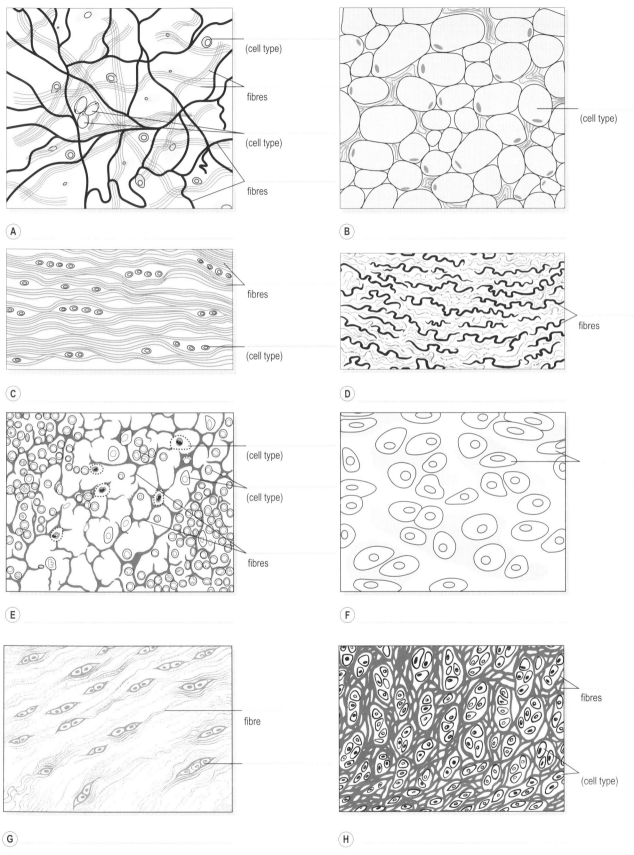

Figure 3.4 Types of connective tissue

Colouring and labelling

25. Name the types of muscle tissue shown in Figure 3.5 and note the differences in the lines below the captions.

26. Colour the nuclei on the muscle tissue shown in Figure 3.5.

27. Label an intercalated disc on Figure 3.5.

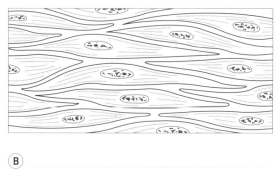

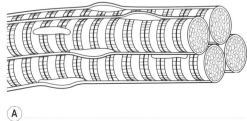

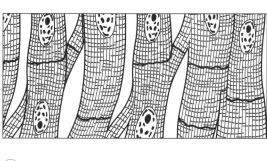

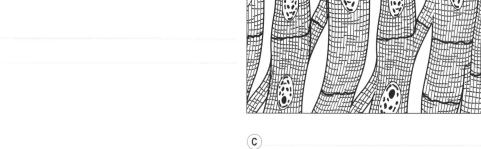

Figure 3.5 Types of muscle tissue

MCQs

28. The epithelial tissue found lining or covering structures subjected to wear and tear is: _____.

    a. Ciliated   **b.** Columnar   **c.** Stratified   **d.** Transitional.

29. Epithelial tissue found on dry body surfaces is relatively waterproof because it contains: _____.

    a. Collagen   **b.** Cilia   **c.** A semisolid matrix   **d.** Keratin.

30. Which of the following are types of connective tissue (choose all that apply): _____.

    a. Adipose tissue   **b.** Blood   **c.** Bone   **d.** Elastic tissue.

31. Which connective tissue cells secrete collagen fibres? _____.

    a. Fibroblasts   **b.** Adipocytes   **c.** Macrophages   **d.** Mast cells.

32. Periosteum, the membrane that covers bones, is formed from: _____.

    a. Elastic tissue   **b.** Fibrous tissue   **c.** Fibrocartilage   **d.** Hyaline cartilage.

33. The intervertebral discs are formed from: _____.

    a. Connective tissue   **b.** Hyaline cartilage   **c.** Fibrocartilage   **d.** Elastic fibrocartilage.

34. The characteristics of cardiac muscle include (choose all that apply): _____.

    a. Branching cells   **b.** Intercalated discs   **c.** Cross stripes   **d.** Spindle-shaped cells.

35. Smooth muscle is found in (choose all that apply): _____.

    a. The heart   **b.** Blood vessel walls   **c.** Ducts of glands   **d.** The urinary bladder.

Completion

36. Complete the blanks in the paragraph below to describe the structure and functions of muscle tissue.

Muscle cells are also called _____. Muscle tissue has the property of _____ that brings about movement, both within the body and of the body itself. This requires a blood supply to provide _____, _____ and _____, and to remove _____. The chemical energy needed is derived from _____.

Skeletal muscle is also known as _____ muscle because _____ is under conscious control. When examined under the microscope, the cells are roughly _____ in shape and may be as long as _____ cm. The cells show a pattern of clearly visible stripes, also known as _____. Skeletal muscle is stimulated by _____ impulses that originate in the brain or spinal cord and end at the _____.

Smooth muscle has the intrinsic ability to _____ and _____, but it can also be stimulated by _____ impulses, some _____ and _____.

Cardiac muscle is found only in the wall of the _____, which has its own _____ system, meaning that this tissue contracts in a co-ordinated manner without external stimulation. _____ impulses and some _____ influence activity of this type of muscle.

37. Complete the paragraphs below to describe characteristics of membranes.

Mucous membrane is sometimes referred to as the _____. It forms the moist lining of body tracts, e.g. the

_____, _____ and _____ tracts. The membrane consists of _____ cells, some of

which produce a sticky secretion called _____. This substance protects the lining from _____. In the

alimentary tract it _____ the contents and in the respiratory system it traps _____.

A serous membrane may also be known as the _____. It consists of a double layer of _____

connective tissue lined by _____ epithelium. The layer lining the body cavity is the _____

layer and that surrounding organs within a cavity, the _____ layer. There are three sites where serous

membranes are found:

   a. the _____ lining the thoracic cavity and surrounding the lungs

   b. the _____ lining the pericardial cavity and surrounding the heart

   c. the _____ lining the abdominal cavity and surrounding the abdominal organs.

Synovial membrane lines the cavities of _____. It consists of _____ tissue containing

_____ fibres. This membrane secretes a clear, sticky, oily substance known as _____.

It provides _____ and _____, and prevents _____ between structures in

_____ joints.

Colouring, matching and labelling

38. Distinguish the exocrine glands on Figure 3.6 by colouring:

   ○ Simple glands
   ○ Compound glands

39. Name the different types of exocrine glands on Figure 3.6.

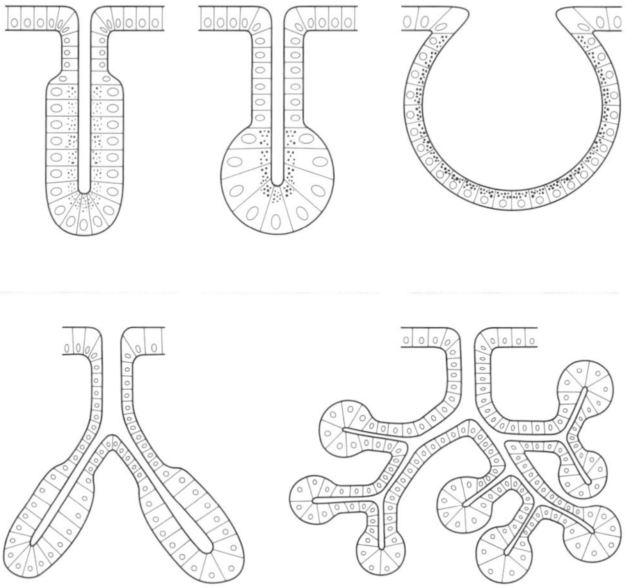

Figure 3.6 Exocrine glands

ORGANIZATION OF THE BODY

 Pot luck

40. The sentence below describes the position assumed in all anatomical descriptions to ensure accuracy and consistency. There are five errors in the sentence. Please correct them to describe the anatomical position.

The body is in the horizontal position with the head facing upwards, the arms facing outwards with the palms of

the hands facing downwards and the feet apart.

Labelling

41. Identify the regional terms indicated on Figure 3.7.

42. Label the directional terms indicated on Figure 3.7.

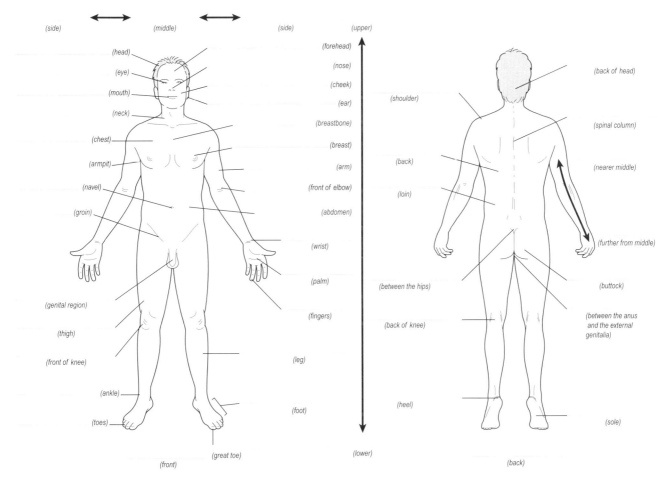

Figure 3.7 Regional and directional terms

Colouring, matching and labelling

43. Colour and match the following parts of the skeleton in Figure 3.8:

○ Axial skeleton
○ Appendicular skeleton

44. Label the bones identified on Figure 3.8.

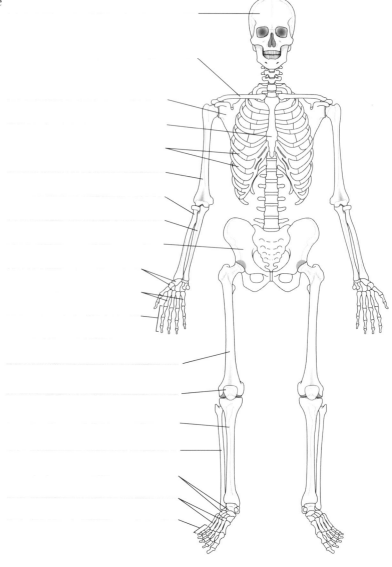

Figure 3.8 The skeleton

 MCQs

45. The anterior bone of the skull is the: _____.

   a. Frontal bone   **b.** Parietal bone   **c.** Temporal bone   **d.** Occipital bone.

46. The inferior border of the thoracic cavity is formed by the: _____.

   a. Rib cage   **b.** Sternum   **c.** Thoracic vertebrae   **d.** Diaphragm.

47. The superior part of the abdominal cavity is formed by the: _____.

   a. Muscles of the abdominal wall   **c.** Diaphragm
   b. Pelvic floor   **d.** Lumbar vertebrae.

48. The mediastinum is found in the: _____.

   a. Pelvic cavity   **b.** Thoracic cavity   **c.** Cranial cavity   **d.** Abdominal cavity.

Colouring and labelling

49. Colour and label the following bones of the cranium and face on Figure 3.9:

Temporal bone	Vomer
Occipital bone	Nasal bone
Parietal bone	Ethmoid bone
Frontal bone	Maxilla
Sphenoid bone	Palatine bone

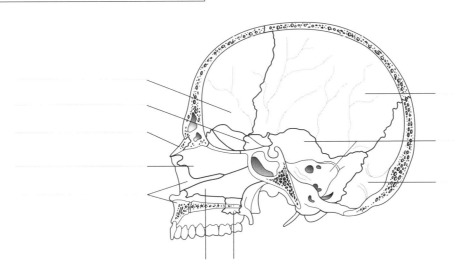

Figure 3.9 The bones forming the cranium and face – viewed from the left

  Colouring and matching

50. Colour and match the following parts of Figure 3.10:

- ○ Intercostal muscles
- ○ Sternocleidomastoid muscles
- ○ Diaphragm
- ○ Ribs
- ○ Vertebrae
- ○ Sternum
- ○ Clavicles
- ○ Costal cartilages

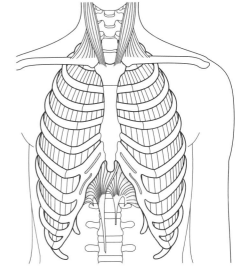

Figure 3.10 Structures forming the walls of the thoracic cavity and associated structures

Pot luck

51. What is the mediastinum? _____.

52. Which accessory digestive organ is situated anteriorly in the abdominal cavity immediately under the diaphragm? _____.

53. Which part of the large intestine lies most anteriorly in the abdominal cavity? _____.

  Colouring, matching and labelling

54. Colour and match the following structures found within the posterior abdominal cavity on Figure 3.11:

- ○ Spleen
- ○ Right adrenal gland
- ○ Kidneys
- ○ Ureters
- ○ Inferior vena cava
- ○ Aorta

55. Colour and label the other organs identified on Figure 3.11.

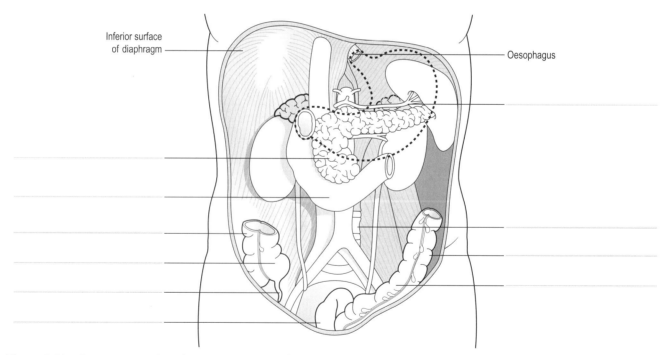

Figure 3.11 Organs occupying the posterior part of the abdominal cavity

Labelling

56. Identify the nine regions of the abdominal cavity shown in Figure 3.12.

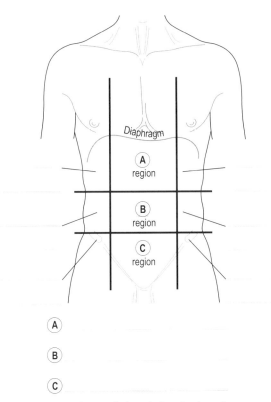

Figure 3.12   Regions of the abdominal cavity

Applying what you know

57. For each organ in list A, state in which body region(s) in List B the organ is situated:

List A

a. Brain: _____

b. Stomach: _____

c. Small intestine: _____

d. Large intestine: _____

e. Lungs: _____

f. Liver: _____

g. Rectum: _____

h. Bladder: _____

i. Heart: _____

List B

1. Hypogastric region
2. Right lumbar region
3. Epigastric region
4. Left iliac region
5. None of these.

58. State which body cavity a surgeon would open to operate on the:

a. Appendix: _____

b. Heart: _____

c. Uterus: _____

d. Stomach: _____

e. Brain: _____

f. Rectum: _____

g. Lungs: _____

h. Spleen: _____

 ## Definitions

Define the following terms:

59. Carcinogen _____

60. Tumour _____

REVISION ACROSS THE CHAPTER

 Matching

61. Match the connective tissue cells in list A to their functions in list B.

List A

1. Plasma cells
2. Adipocytes
3. Fibroblasts
4. Mast cells
5. Macrophages.

List B

a. Secrete collagen fibres _____

b. Storage of fat _____

c. Phagocytes that digest cell debris and invading bacteria _____

d. Synthesis and secrete antibodies _____

e. Contain granules that release histamine and heparin at sites of tissue damage

_____ .

 Pot luck

62. Cross out the incorrect options to use the directional terms correctly.

   a. The humerus is **medial/lateral** to the heart

   b. The vertebrae are **anterior/posterior** to the kidneys

   c. The phalanges are **proximal/distal** to the ulna

   d. The skull is **inferior/superior** to the vertebral column

   e. The greater omentum is **anterior/posterior** to the small intestine

   f. The appendix is **inferior/superior** to the stomach

   g. The patella is **proximal/distal** to the tarsal bones

   h. The scapulae are **medial/lateral** to the sternum

   i. The kidneys are **inferior/superior** to the adrenal glands.

 MCQs

63. The correct order of the stages of mitosis is: _____.

   a. Anaphase, metaphase, prophase, telophase
   b. Prophase, metaphase, anaphase, telophase
   c. Telophase, metaphase, prophase, anaphase
   d. Prophase, anaphase, telophase, metaphase.

64. Endocrine glands secrete: _____.

   a. Enzymes   **b.** Hormones   **c.** Mucus   **d.** Serous fluid.

65. The main function of the intercostal muscles is in: _____.

   a. Swallowing   **b.** Breathing   **c.** Walking   **d.** Extending the neck.

66. The bones of the vertebral column, from above downwards are: _____.

   a. 5 cervical, 10 thoracic, 6 lumbar, coccyx (fused), sacrum (fused)
   b. 7 cervical, 10 thoracic, 5 lumbar, sacrum (fused), coccyx (fused)
   c. 10 cervical, 12 thoracic, 6 lumbar, sacrum (fused), coccyx (fused)
   d. 7 cervical, 12 thoracic, 5 lumbar, sacrum (fused), coccyx (fused).

The blood

The blood is a fluid connective tissue, which travels within the closed circulatory system. It carries nutrients, wastes, respiratory gases and other substances important to body function. This chapter will test your understanding of the physiology of blood.

COMPOSITION OF BLOOD

 Labelling and colouring

1. Figure 4.1A shows whole blood that has been prevented from clotting and allowed to stand for some time. Label and colour the two layers shown.

2. Figure 4.1B shows whole blood that has been allowed to clot. Label and colour the parts shown.

3. What is present in the fluid portion in Figure 4.1A that is absent in Figure 4.1B? _____.

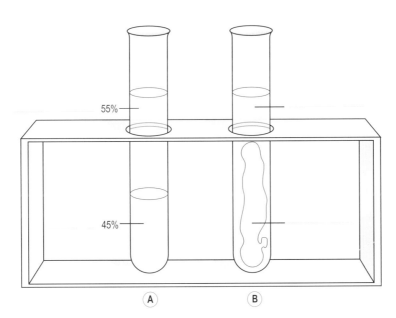

55%

45%

A   B

Figure 4.1 The proportions of blood cells and plasma in whole blood separated by gravity. **A.** Blood prevented from clotting. **B.** Blood allowed to clot

Matching

4. List A presents various descriptive phrases. Match them with the suggested components of plasma in the box of key choices. (Be careful, not every key choice is actually found in plasma!)

List A

a. Protective proteins of the immune system _____

b. Mineral required for healthy bones and teeth _____

c. Secreted by the endocrine system _____

d. A clotting factor synthesized by the liver _____

e. The major constituent of plasma _____

f. Preferred fuel source for body cells _____

g. A waste product of protein breakdown _____

h. Carbon dioxide is carried in this form _____

i. Deficiency of this substance can cause anaemia _____

j. Transport steroid hormones in blood _____ .

> *Key choices:*
> Urea   Hormones
> Haemoglobin   Water
> Amino acids   Albumin
> Fibrinogen   Bile
> Glucose   Antibodies
> Rhesus antigens   Bicarbonate ion
> Intrinsic factor   Iron
> Calcium

5. This question considers the characteristics and functions of plasma proteins. Match each item in list A to the appropriate key choices listed below. You may use each key choice more than once.

List A

a. Synthesized in the liver _____

b. Principally responsible for maintaining plasma osmotic pressure _____

c. Present in blood in its inactive form _____

d. The two main plasma proteins contributing to blood viscosity _____

e. Synthesized in lymphoid tissue _____

f. Carrier molecule for lipids in blood _____

g. Carrier molecule for thyroxine _____

h. Present in plasma but absent from serum _____ .

> *Key choices:*
> Immunoglobulins
> Thyroglobulin
> Fibrinogen
> Albumins

CELLULAR CONTENT OF BLOOD

  Colouring and labelling

6. Figure 4.2 shows the eight main types of blood cell. Name each type in the space provided.

7. In Figure 4.2, colour the granules in the cytoplasm of those white cells that contain them and the nuclei of the cells that have them.

8. The term used to describe blood cell formation is:

_____ .

 Pot luck

9. Explain why cell (A) has no nucleus.

_____ .

10. List three substances that are found within the cytoplasmic granules of cells (C), (D) and (E).

_____ .

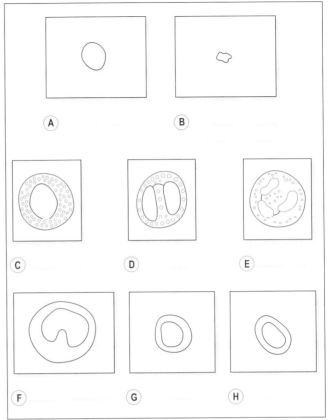

Figure 4.2 Blood cells

 Matching

11. The letters in list A correspond with the blood cells in Figure 4.2. The box below lists 20 numbered key choices that can be used to describe each of them. Match each cell type with the relevant key choices. (Be careful, you can use each key choice more than once!)

List A
List the numbers of the key choices here.

a: _____

b: _____

c: _____

d: _____

e: _____

f: _____

g: _____

h: _____

Key choices:
1. Circulating mast cell
2. Makes antibodies
3. Important in clotting
4. Granulocyte
5. Agranulocyte
6. Commonest blood phagocyte
7. 1–6% of total white blood cells
8. Cell fragment
9. Large single nucleus
10. Involved in immunity
11. Diameter of about 7 microns
12. Has no nucleus
13. Contain haemoglobin
14. 0.04–0.44×10^9 cells/litre
15. Smallest white blood cell(s)
16. 2–10% of total white cells
17. Made in red bone marrow
18. Originate from pluripotent stem cells
19. Synthesis is called erythropoiesis
20. Biconcave in shape

MCQs

12. What is the average lifespan of a red blood cell? _____.

    a. 3 weeks   **b.** 3 months   **c.** 6 weeks   **d.** 6 days.

13. Which two substances are necessary for normal red blood cell maturation? _____.

    a. Intrinsic factor and folic acid   **c.** Folic acid and iron
    b. Iron and vitamin B_{12} **d.** Folic acid and vitamin B_{12}.

14. Where in the body is the equation $Hb + O_2 \rightleftharpoons HbO_2$ driven to the right? _____.

    a. In the lungs   **b.** In the kidneys   **c.** In the heart   **d.** In the brain.

15. Which of the following describes the structure of a haemoglobin molecule? _____.

    a. Four protein chains and one iron atom   **c.** Four haem groups and four protein chains
    b. Four haem groups and one iron atom   **d.** Four iron atoms and one haem group.

16. If a haemoglobin molecule is saturated, which of the following would be true? _____.

    a. All six of its oxygen binding sites are full
    b. The haemoglobin molecule is carrying its full complement of iron
    c. It has collected carbon dioxide in the tissues and has changed from bright red to bluish in colour
    d. The molecule is likely to be in the pulmonary vein rather than in a systemic vein.

17. Which of the following would increase the release of oxygen from oxyhaemoglobin? _____.

    a. Decreased tissue metabolism   **c.** Decreased tissue carbon dioxide production
    b. Increased tissue temperature   **d.** Increased red blood cell numbers.

 Matching

18. Figure 4.3 is a flow chart describing red blood cell synthesis. Complete it by putting the statements below into the diagram in the right order.

> Bone marrow increases erythropoiesis
>
> Increased blood oxygen-carrying capacity reverses tissue hypoxia
>
> Red blood cell numbers rise
>
> Kidneys secrete erythropoietin into the blood

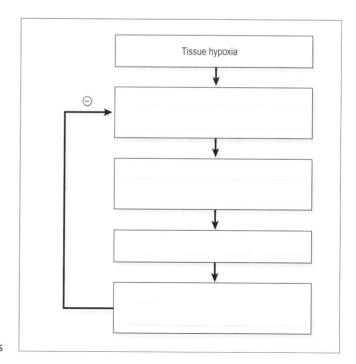

Figure 4.3 Control of erythropoiesis

Completion

19. The following paragraph describes the destruction of red blood cells. Complete it by filling in the blanks.

The lifespan of red blood cells is usually about _____ days. Their breakdown, also called

_____, is carried out by phagocytic _____ cells found mainly in the

_____, _____ and _____. Their breakdown releases the mineral

_____, which is kept by the body and stored in the _____. It is used to form new

_____. The protein released is converted to the intermediate _____, and then to the

yellow pigment _____, before being bound to plasma protein and transported to the

_____, where it is excreted in the _____.

20. Complete Table 4.1, which describes the ABO system of blood grouping.

Blood group	Type of antigen present on red cell surface	Type of antibody present in plasma	Can safely donate to:	Can safely receive from:
A				
B				
AB				
O				

Table 4.1 The ABO system of blood grouping

21. Which of the blood groups in Table 4.1 is known as the universal donor? _____.

22. Which of these blood groups is known as the universal recipient? _____.

Applying what you know

23. Figure 4.4 shows the results of a blood typing test. Blood samples from Harold, Olivia, Alex and Amanda have been mixed with anti-A antibodies (column A), anti-B antibodies (column B) or both anti-A and anti-B antibodies (column AB). The symbol ✳ indicates that a reaction has taken place between their blood and the antibodies added. For each individual, work out their blood group and enter it in the far right hand column.

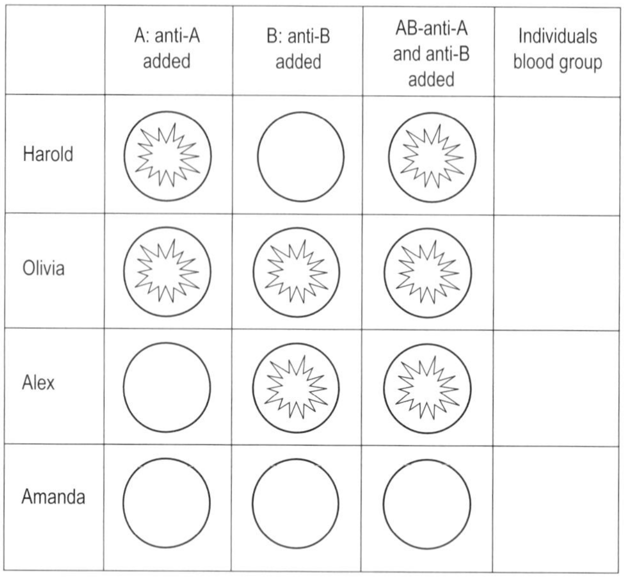

	A: anti-A added	B: anti-B added	AB-anti-A and anti-B added	Individuals blood group
Harold	✳	◯	✳	
Olivia	✳	✳	✳	
Alex	◯	✳	✳	
Amanda	◯	◯	◯	

Figure 4.4 The results of a blood typing test for four individuals with different blood groups.

24. Using the results seen in Figure 4.4, which individuals could possibly donate safely to Harold? Explain your reasoning.

_____

_____

_____

25. Complete Table 4.2, which describes the function of the main white blood cells, by ticking the appropriate boxes against each cell type.

	Neutrophils	Eosinophils	Basophils	Monocytes	Lymphocytes
Phagocyte					
Involved in allergy					
Converted to macrophages					
Release histamine					
Many in lymph nodes					
Kupffer cells					
Increased numbers in infections					
Kill parasites					
Part of the reticuloendothelial system					

Table 4.2   Characteristics of white blood cells

 Completion

26. Figure 4.5 shows the basic stages of blood clotting. Complete the diagram by filling in the blanks using the key choices in the box.

Key choices:
Fibrin clot   Thrombin
Intrinsic   Final common pathway
Platelets   Extrinsic
Prothrombin   Blood vessel lining
Thromboplastin   Fibrinogen
Loose fibrin

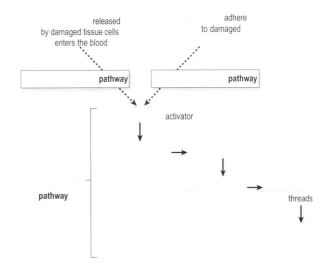

Figure 4.5   Stages of blood clotting

Pot luck

27. The following statements refer to blood clotting. Two are correct but six contain errors. Identify the correct statements and correct the errors.

   a. The intrinsic pathway is activated by platelets adhering to damaged blood vessel endothelium.

   _____

   _____

   b. The extrinsic and intrinsic pathways both activate the first step in the final common pathway, which is the conversion of fibrinogen to fibrin.

   _____

   _____

   c. Clotting factors circulate in the bloodstream in an active form, so that clotting occurs rapidly when required.

   _____

   _____

   d. Prothrombin is clotting factor IV.

   _____

   _____

   e. Thrombin is the principal enzyme involved in the breakdown of clots once bleeding has been stopped.

   _____

   _____

   f. Although the final common pathway is triggered by both the extrinsic and intrinsic pathways, the extrinsic pathway is the faster of the two.

   _____

   _____

   g. The platelet plug formed rapidly following blood vessel damage is a sturdy structure because it contains large quantities of fibrin.

   _____

   _____

   h. Blood clotting is an example of a negative feedback mechanism, since otherwise blood would progressively clot within the vessels even when not required.

   _____

   _____

The cardiovascular system

The cardiovascular system consists of the heart, which is a pump, and the vast network of vessels, which are the transport system for the blood. Together, they supply all the body's tissues with nutrients and carry away wastes. This chapter will help you to understand its structure, the functions of the heart and the different types of blood vessel, and the control of blood pressure. The lymphatic system, which is also important in fluid transport, is dealt with in a separate chapter.

BLOOD VESSELS

 Completion

1. Complete the following paragraph, which describes the two circulation systems of the blood, by inserting the correct word in the spaces provided.

The heart pumps blood into two separate circulatory systems, the _____ circulation and the

_____ circulation. The _____ side of the heart pumps blood to the lungs, whereas

the _____ side of the heart supplies the rest of the body. The _____ are the sites of exchange of

nutrients, gases and wastes. Tissue wastes, including carbon dioxide, pass into the _____ and the tissues

are supplied with _____ and _____.

 Labelling and colouring

2. Label and colour Figure 5.1.

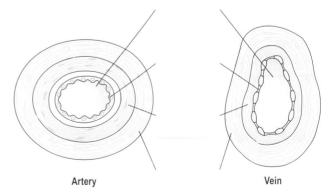

Artery   Vein

Figure 5.1 Structures of an artery and a vein

Completion

3. Match each of the three layers you identified in Figure 5.1 with the two most appropriate descriptive phrases given in Table 5.1.

Descriptive phrase	Layer (tunica) of vessel wall
Squamous epithelium	
Contains mainly fibrous tissue	
Endothelial layer	
Consists partly of muscle tissue	
The vessel's elastic tissue is here	
Outer layer	

Table 5.1 Layers of vessel wall

? MCQs

4. Which of the following describes the correct sequence of blood vessels through which blood flows after leaving the heart? _____.

    a. Veins, venules, capillaries, arterioles, arteries
    b. Capillaries, venules, veins, arteries, arterioles
    c. Arteries, arterioles, capillaries, venules, veins
    d. Arterioles, capillaries, venules, veins, arteries.

5. Which of the following relates to capacitance vessels? _____.

    a. These are the arterioles, because they have the capacity to control blood pressure
    b. The capillaries are capacitance vessels, because they are so numerous they contain most of the body's blood at any one time
    c. Any large-diameter vessel, artery or vein can be described as a capacitance vessel
    d. Veins are capacitance vessels, because they are distensible and can widen to hold more blood.

6. Arteries have thicker walls than veins because they: _____.

    a. carry less blood than veins
    b. carry blood at higher pressure than veins
    c. have to provide support for the valves in their lumens
    d. have a much thicker endothelium than veins.

7. The tiny arteries that link one artery to another are called: _____.

    a. muscular arteries   b. link arteries   c. anastomoses   d. resistance vessels.

8. The thickest layer of tissue in the wall of the aorta is: _____.

    a. elastic tissue   b. smooth muscle   c. fibrous supporting tissue   d. protective adipose tissue.

Completion

9. Complete the following paragraph by inserting the correct term in the blanks provided.

The smallest arterioles split up into a large number of tinier vessels called _____. Across the walls

of these vessels, the tissues obtain _____ and _____, and get rid of their

_____. The walls of these vessels are therefore thin, being only _____ thick.

Substances such as _____ and _____ can pass across them, whereas larger

constituents of blood such as _____ and _____ are retained within the vessel. This

vast network of microscopic vessels has a diameter of only about _____, and links the arterioles to

the _____. In some parts of the body, such as the liver, the vessels in the tissues are wider than this,

and are called _____. Blood flow here is _____ than in other tissues because of the

bigger lumen.

? **Pot luck**

The following questions relate to regulation of blood flow.

10. What effect does sympathetic stimulation have on blood vessel diameter in most systemic beds?

_____.

11. Why does sympathetic stimulation have less effect on blood vessels in the brain than on vessels in the skin?

_____.

12. Identify the three factors that determine resistance in a blood vessel:

- _____

- _____

- _____.

13. Which of these three is most important in controlling peripheral resistance in blood vessels?

_____.

Completion

14. The following paragraph relates to autoregulation. Complete by crossing out the incorrect options from the choices in bold.

Autoregulation is the **ability of a tissue to control its own blood flow/automatic control of blood flow by the cardiovascular centre/effect of the autonomic nervous system on blood vessel diameter**. It is regulated **by the brain/by circulating hormones in the blood/locally**. Two examples of autoregulation are **the increase of blood flow through the gastrointestinal tract after eating/the rise in insulin levels in the blood after eating/the reduction in blood flow through the skin when cold**. Another example is the **increase/decrease** in blood flow through an active tissue such as exercising skeletal muscle. Blood vessels supplying an active muscle are **constricted/dilated** by **reduced temperature/hypoxia/low CO$_2$ levels** in the tissue, and this leads to an **increased/decreased** blood supply to match tissue needs.

Matching

15. Substances moving in and out of capillaries do so usually by one of the following processes: osmosis, diffusion or active transport. Decide whether the following statements apply to any one or any combination of these three, and complete Table 5.2 by ticking in the appropriate columns.

	Osmosis	Diffusion	Active transport
Movement only down a concentration gradient			
Movement of water molecules			
Movement across a semipermeable membrane			
Movement requires energy			
Movement up a concentration gradient possible			
Movement does not require energy			
Movement of oxygen			
Movement of carbon dioxide			

Table 5.2 Characteristics of osmosis, diffusion and active transport

16. For each of the five statements in list A, identify the most appropriate answer from list B. (You may need the items in list B more than once.)

List A

a. Plasma proteins in the bloodstream are responsible for exerting _____

b. The hydrostatic pressure of the blood is also known as _____

c. The main force opposing the osmotic pressure of the blood is _____

d. The main force pushing fluid out of the arterial end of the capillary is the _____

e. The main force drawing fluid back into the venous end of the capillary is the _____.

List B

Hydrostatic pressure
Blood pressure
Osmotic pressure

  Matching and colouring

17. Figure 5.2 shows the effect of capillary pressures on water movement in and out of the capillary. Using different colours, colour and match the arrows to represent hydrostatic and osmotic pressures at each end of the capillary.

- ○ Hydrostatic pressure
- ○ Osmotic pressure

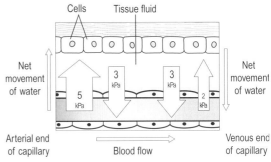

Figure 5.2 Effect of capillary pressures on water movement between capillaries and cells

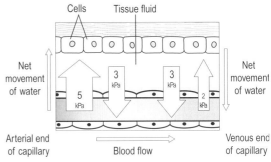 **Pot luck**

18. Explain why hydrostatic pressure is greater at the arterial end of the capillary than at the venous end.

_____

_____

19. This pressure difference leads to an overall accumulation of fluid in the tissues. What happens to this extra fluid?

_____

_____

THE HEART

  Colouring and labelling

20. Label Figure 5.3.

21. On Figure 5.3, colour the vessels carrying oxygenated blood red and the vessels carrying deoxygenated blood blue.

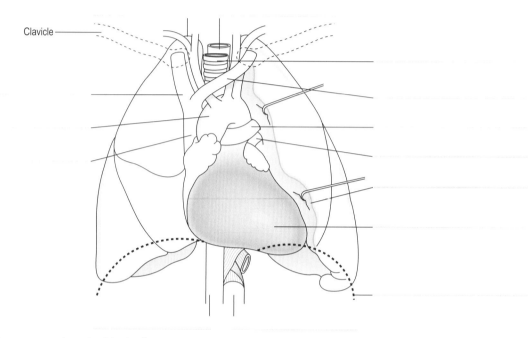

Clavicle

Figure 5.3 Organs associated with the heart

  Colouring and labelling

22. Figure 5.4 shows the main layers of the heart wall. Colour and label the structures shown.

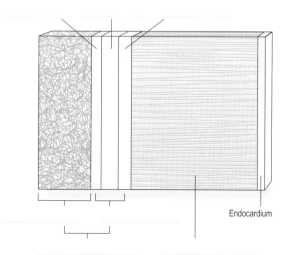

Endocardium

Figure 5.4 Layers of the heart wall

Matching

23. Match the following layers of the heart wall to the statements in Table 5.3. You may use each layer more than once, but there is only one correct answer for each statement.

Myocardium	Serous pericardium	Endocardium	Pericardial space
Parietal pericardium	Visceral pericardium	Fibrous pericardium	

Statement	Layer
Lines the heart chambers	
Consists of cardiac muscle	
One cell thick	
Lines the fibrous pericardium	
Outer supportive layer	
Secretes pericardial fluid	
Contains pericardial fluid	
Lies between the visceral and serous pericardial layers	
Lies between the endocardium and the visceral pericardium	
The thickest layer of the heart wall	
Inelastic layer	
Covers the heart valves	
Covers the myocardium	

Table 5.3   Features of heart wall structure

? **MCQs**

24. Cardiac muscle cells: _____.
    a. are found in the heart and in the walls of blood vessels
    b. are striated, like skeletal muscle cells
    c. are larger than skeletal muscle cells
    d. contain more than one nucleus.

25. The myocardium is: _____.
    a. thickest at the apex of the heart
    b. lined with fine fibrous tissue
    c. the main tissue found in heart valves
    d. under voluntary control.

26. Regarding electrical stimulation of the myocardium: _____.
    a. each cardiac muscle cell is supplied by its own nerve ending
    b. electrical activity spreads from one cell to the next via intercalated discs
    c. fibrous tissue in the heart helps to spread the electrical activity
    d. each cardiac muscle cell generates its own electrical signal.

27. What proportion of the cardiac output goes directly to the myocardium? _____.
    a. 5%   b. 8%   c. 12%   d. 15%.

 Pot luck

28. The double membrane arrangement of the serous pericardium is found in which two other locations in the body?

    • _____   • _____.

Labelling

29. Name the chambers of the heart shown on Figure 5.5.

A:	B:
C:	D:

30. Label all the structures indicated on Figure 5.5.

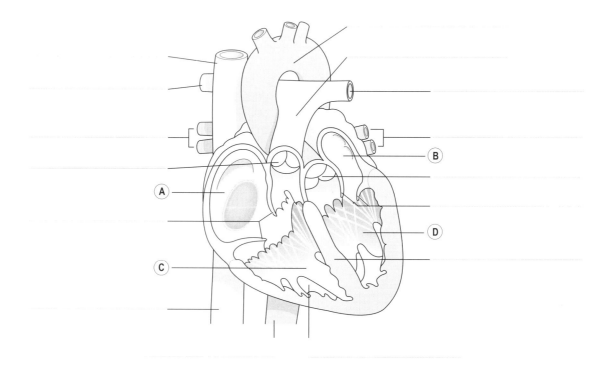

Figure 5.5   Interior of the heart

Pot luck

31. What is the function of the chordae tendineae?

_____

_____

_____ .

32. What is the function of the valves?

_____

_____

_____ .

 Labelling

33. Label the structures indicated in Figure 5.6.

34. Using red arrows, indicate the direction of flow of oxygenated blood through the appropriate vessels and chambers; using blue arrows, do the same for deoxygenated blood.

 Matching

35. Put the terms supplied below in the correct order, so that they correctly describe the flow of blood through the pulmonary and systemic circulations, beginning and finishing with the aorta.

Left atrium
Right atrioventricular (tricuspid) valve
Right atrium
Systemic arterial network
Systemic venous network
Left ventricle
Pulmonary valve
Left atrioventricular (mitral) valve
Right ventricle
Pulmonary arteries
Capillaries of body tissues
Aortic valve
Pulmonary veins

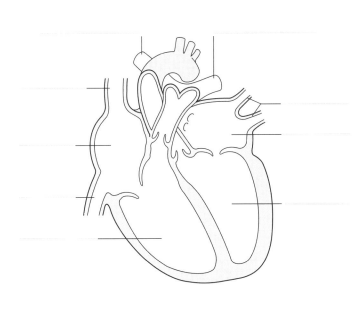

Figure 5.6 Direction of blood flow through the heart

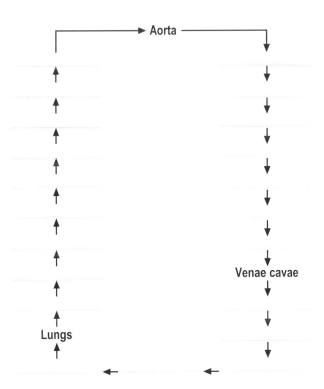

36. Explain why the muscle layer in the left ventricle is thicker than in the right ventricle.

_____

_____

_____

_____ .

  Matching and colouring

37. In Figure 5.7, identify and label the left and right sides of the heart, their respective valves, the pulmonary circulation and the systemic circulation and label the capillary beds representing the lungs and the body tissues.

38. Colour the arrows on the diagram representing transport of oxygenated blood red, and arrows showing transport of deoxygenated blood blue.

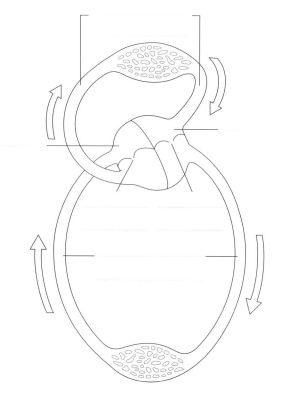

Figure 5.7 Relationship between the pulmonary and systemic circulations

  Colouring and labelling

39. Figure 5.8 shows the main blood vessels associated with the heart and the vessels supplying the heart muscle. Colour and label vessels E and F.

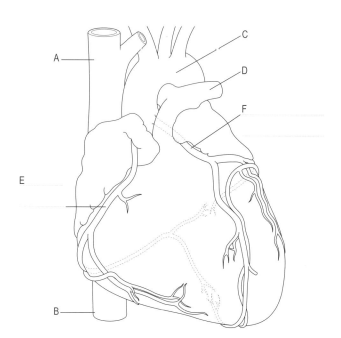

Figure 5.8 Blood supply to the heart

40. What is the function of vessels E and F?

_____.

41. From which of the vessels A, B, C or D do vessels E and F arise? Name and colour the correct vessel.

_____.

42. Which condition is likely to occur if either vessels E or F, or one of their major branches, becomes blocked?

_____.

43. Label the structures indicated on Figure 5.9, which shows the conducting system of the heart, and colour the conducting tissue.

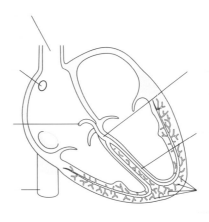

Figure 5.9 Conduction system of the heart

? MCQs

44. The heart rate is regulated by the cardiovascular centre, which lies where in the brain? _____.

    a. In the cerebellum
    b. In the thalamus
    c. In the medulla oblongata
    d. In the cerebrum.

45. Which of the following lists three effects that will all increase heart rate? _____.

    a. Sympathetic activation; active exercise; fear
    b. Adrenaline (epinephrine) release; reduced exercise; fall in blood pressure
    c. Parasympathetic stimulation; rise in blood pressure; thyroxine release
    d. Fall in blood pressure; adrenaline (epinephrine) release; sleep.

46. Which of the following lists three effects that will all decrease heart rate? _____.

    a. Parasympathetic activation; thyroxine release; increase in blood pressure
    b. Adrenaline (epinephrine) release; sympathetic inhibition; reduced exercise
    c. Increase in blood pressure; sympathetic inhibition; sleep
    d. Fear; parasympathetic stimulation; sympathetic inhibition.

47. Which of the following statements is true? _____.

    a. Vagal stimulation increases heart rate
    b. The sinoatrial node is supplied by both sympathetic and parasympathetic fibres
    c. The neurotransmitter released by parasympathetic fibres at heart muscle is noradrenaline
    d. Heart rate rises when acetylcholine acts on heart muscle cells.

Completion

48. Complete the following paragraphs, which describe the cardiac cycle.

Diastole

We will begin this description with the heart in diastole, when the whole heart is _____. During

this time, in the upper part of the heart, the atria are _____ and blood is flowing

_____. Not only the upper chambers are filling but also the lower ones; because the

_____ valves are open, we see that the _____ are filling as well. Although blood is

travelling into the lower chambers, at this stage the electrical activity has not reached them yet and so the

ventricles are _____. Remember, during this period, the heart muscle is not contracting; both the

_____ and the _____ are relaxed.

Atrial systole

The next stage represents atrial systole, or contraction. This is initiated when the _____ fires; its

electrical discharge leads to the spread of _____ through the atria. Because of the electrical

excitation of the muscle, the atria _____ and this leads to pumping of blood from the

_____ into the _____. It is important therefore that the _____ valves

are open, to permit blood to flow through. The ventricles fill up; because the _____ and

_____ valves are closed, blood cannot yet pass from the heart into the great vessels leaving it.

Ventricular systole

The third stage is ventricular systole. The impulse from the sinoatrial node has passed through the atrioventricular

node; inspection of the atria shows that they are _____ after their period of activity; this allows

them to rest. However, as far as the lower chambers are concerned, because _____ are spreading

through the ventricles, we see that the ventricles _____. So that blood cannot flow in a backwards

manner into the atria, the _____ valves are closed. However, for the ventricles to be able to push

blood out of the heart, the _____ and _____ valves _____. Because

of the force generated by the contracting ventricular muscle, blood is pumped from the ventricles into the

_____ and the _____.

   The cycle is now complete; the heart will enter another period of diastole, allowing the entire organ to rest

briefly before the next period of contraction.

Pot luck

49. In each of the sentences related to the cardiac cycle in the box below, the initial statement is true. In four of the sentences, the reason given does not explain the initial statement. Identify the four inaccurate reasons and correct the explanation.

	Initial statement		Reason
A.	When the ventricles contract, the atrioventricular valves close,	because	the muscle in the valves relaxes and allows them to close.
B.	During ventricular contraction, the aortic valve is open,	because	pressure in the aorta is high.
C.	The stages of each heartbeat (cardiac cycle) can be monitored with a stethoscope placed on the chest,	because	when the heart valves close, they generate characteristic sounds.
D.	When the atria contract, the atrioventricular valves open,	because	the pressure in the atria is greater than the pressure in the ventricles.
E.	There is a very brief delay between atrial contraction and ventricular contraction,	because	transmission of the electrical signal slows down very slightly as it passes through the sinoatrial node.
F.	Blood does not flow backwards in the heart, i.e from ventricles to atria,	because	the pressure in the ventricles is never high enough to force blood backwards.

A. _____

B. _____

C. _____

D. _____

E. _____

F. _____

 Labelling

50. Figure 5.10 shows a typical electrocardiograph of one cardiac cycle. Label the individual waves.

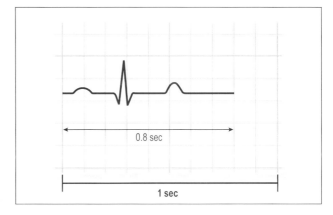

Figure 5.10 The electrocardiogram (ECG)

Matching

51. Match each of the ECG waves named in Figure 5.10 with the corresponding electrical and mechanical events, listed below.

   a. Passage of electrical impulse through the Purkinje fibres: _____

   b. Atrial systole: _____

   c. Ventricular relaxation: _____

   d. Opening of the aortic and pulmonary valves: _____

   e. Discharge of the sinoatrial node: _____

   f. Ventricular systole: _____

   g. Atrial relaxation: _____

   h. Closure of the atrioventricular valves: _____.

Applying what you know

52. In a man whose heart rate is 80 beats per minute and whose stroke volume is 70 ml, what is the cardiac output? _____.

53. In a woman whose cardiac output is 6 L/min and whose pulse rate is 75 beats per minute, what is the stroke volume? _____.

54. In a man whose cardiac output is 5 L/min and whose stroke volume is 50 ml, what is the heart rate? _____.

55. If each cycle on an ECG trace lasts 0.5 sec, what is the heart rate? _____.

56. If each cycle on an ECG trace lasts 1.2 sec, what term is used to describe the heart rate? _____.

Pot luck

57. Which of the following would increase stroke volume (assuming no other factor changes to compensate)? Tick all that apply.

   a. Sympathetic stimulation: _____   **e.** Decreased secretion of adrenaline (epinephrine): _____

   b. Increased preload: _____   f. Decreased afterload: _____

   c. Increased vagal tone: _____   **g.** Increased blood volume: _____

   d. Increased heart rate: _____   h. Decreased venous return: _____.

58. Which of the following is associated with increased venous return to the heart (assuming no other factor changes to compensate)? Tick all that apply.

a. Standing up from a supine position: _____

b. Decreased blood volume: _____

c. The skeletal muscle pump: _____

d. Increased blood pressure: _____

e. Decreased heart rate: _____

f. The respiratory pump: _____

g. Venous congestion: _____

h. Increased preload: _____.

BLOOD PRESSURE

 MCQs

59. Systolic blood pressure represents which stage of the cardiac cycle?

a. Ventricular diastole

b. The end of the cardiac cycle

c. Contraction of the ventricles

d. Relaxation of both atria and ventricles.

60. Which equation determines blood pressure?

a. Heart rate × cardiac output

b. Peripheral resistance × cardiac output

c. Peripheral resistance × pulse pressure

d. Heart rate × pulse pressure.

61. Peripheral resistance is determined mainly by:

a. arteriolar diameter

b. venous return

c. heart rate

d. pulse pressure.

62. Why does blood pressure not fall to zero when the heart is at rest?

a. Blood is constantly moving in the body's blood vessels

b. The heart is never at rest

c. Blood vessels have resting tone

d. There is more blood in the arterial system than in the venous system, so pressure is higher.

Completion

63. Complete the following paragraphs, which describe the body's control of blood pressure, by scoring out the incorrect options in bold, thus leaving the correct words or phrases.

The baroreceptor reflex is important in the **moment-to-moment/long-term** control of blood pressure. It is controlled by the cardiovascular centre found in the **medulla oblongata/carotid bodies**, and which receives and integrates information from baroreceptors, chemoreceptors and higher centres in the brain. Baroreceptors are receptors sensitive to blood pressure and are found in the **carotid arteries/heart wall/aorta**. A **rise/fall** in blood pressure activates these receptors, which respond by increasing the activity of **parasympathetic/sympathetic** nerve fibres supplying the heart; this **slows the heart down/speeds the heart up** and returns the system towards normal. In addition to this, **sympathetic/parasympathetic** nerve fibres supplying the blood vessels are **activated/inhibited**, which leads to **vasoconstriction/vasodilation**, again returning the system towards normal (note that most blood vessels have little or no **sympathetic/parasympathetic** innervation).

On the other hand, if the blood pressure **falls/rises**, baroreceptor activity is decreased, and this also triggers compensatory mechanisms. This time, **sympathetic/parasympathetic** activity is increased and this leads to a(n) **reduction/increase** in heart rate; in addition, cardiac contractile force is **increased/reduced**. The blood vessels respond with **vasoconstriction/vasodilation**; this is mainly due to **increased/decreased** activity in **sympathetic/ parasympathetic** fibres. These measures lead to a restoration of blood pressure towards normal.

In addition to the activity of the baroreceptors described above, chemoreceptors in the **carotid bodies/aorta/ higher centres of the brain** measure the pH of the blood. Increase in **oxygen/carbon dioxide** content of the blood decreases pH and **stimulates/inhibits** these receptors, leading to an **increase/decrease** in stroke volume and heart rate, and a general **vasoconstriction/vasodilation**; this **increases/decreases** blood pressure. Other control mechanisms include the renin–angiotensin system, which is involved in **long-term/short-term** regulation; activation **increases/decreases** blood volume, thereby **increasing/decreasing** blood pressure.

? **Pot luck**

64. Where are the main chemoreceptors responsible for regulation of respiration:

- outwith the CNS?   - within the CNS?

- _____   - _____ .

- _____

65. Identify three chemical changes that stimulate the chemoreceptors that control respiration.

- _____   - _____   - _____ .

CIRCULATION OF THE BLOOD

  Colouring and matching

66. On Figure 5.11, indicate the locations of the main pulse points by using different colours in the key.

○ Temporal artery
○ Carotid artery
○ Facial artery
○ Femoral artery
○ Brachial artery
○ Popliteal artery
○ Dorsalis pedis artery
○ Posterior tibial artery
○ Radial artery

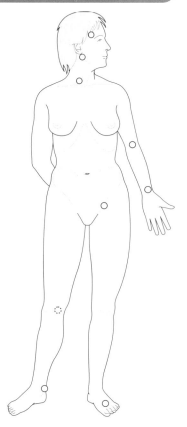

Figure 5.11 Main pulse points

 Completion

67. The following paragraph describes the flow of blood through the pulmonary circulation. Complete it by filling in the blanks.

Blood leaving the right ventricle first enters the _____, which passes upwards close to the aorta

and divides into the right _____ and the left _____ at the level of the 5th thoracic

vertebra. Each of these branches goes to the corresponding _____, and enters these organs in the

area called the _____. Within the tissues, the vessels divide and subdivide, giving a network of

many millions of tiny _____, across the walls of which gases exchange. Blood draining these

structures then passes through veins of increasing diameter, which finally unite in the _____,

which carry the blood back to the _____ atrium of the heart.

Matching

68. The artery leaving the heart and entering the systemic circulation is the aorta (see Figure 5.12), which travels behind the heart, penetrates the diaphragm and descends into the abdomen. Label its main parts and branches using the key choices listed below (L/R = left/right; A = artery).

L. internal iliac A.	Coeliac A.
Inferior mesenteric A.	Ascending aorta
Abdominal aorta	L. common iliac A.
Brachiocephalic A.	R. renal A.
L. subclavian A.	R. subclavian A.
R. common carotid A.	Superior mesenteric A.
Arch of aorta	L. common carotid A.
L. external iliac A.	Thoracic aorta

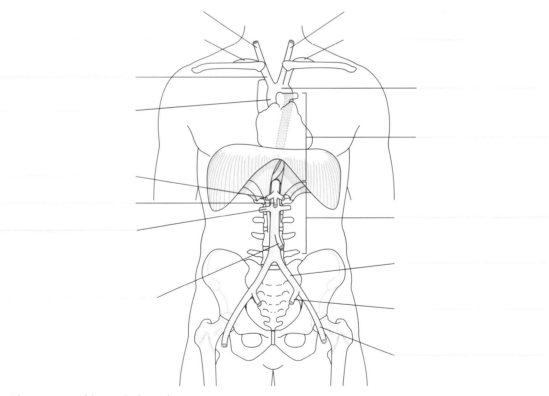

Figure 5.12 The aorta and its main branches

69. Match the statements in Table 5.4 to the appropriate artery in list A.

Statement	Artery
Main supplier to the circulus arteriosus	
Supplies the superficial tissues of the head and neck	
Carotid sinuses occur where this artery bifurcates	
Excepting the coronary arteries, this is the first artery to branch from the arch of the aorta	
One of the contributing arteries to the circulus arteriosus	
This artery can be felt as a pulse point just in front of and above the ear	

Table 5.4 Features of some important arteries supplying the head and neck

List A

Temporal artery   Internal carotid artery
Common carotid artery   Brachiocephalic artery
Basilar artery   External carotid artery

⊡ Labelling

70. Figure 5.13 shows the circulus arteriosus (circle of Willis), which is important in supplying most of the brain. Label the arteries indicated.

71. Venous blood from deep areas of the brain is collected in channels called sinuses, which empty ultimately into the internal jugular veins. Figure 5.14 shows the main venous sinuses of the left side of the brain (remember a mirror image will also exist on the right hand side). Label the sinuses indicated.

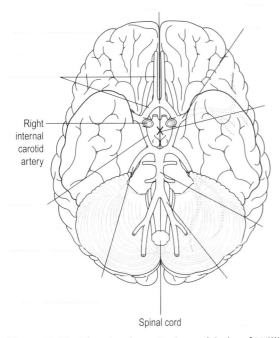

Figure 5.13 The circulus arteriosus (circle of Willis)

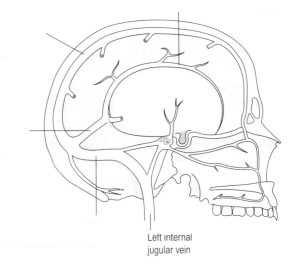

Figure 5.14 Venous sinuses of the brain

Labelling and colouring

72. Figures 5.15 and 5.16 show the main arteries and veins of the limbs. Label the vessels shown, and colour the arteries in red and the veins in blue.

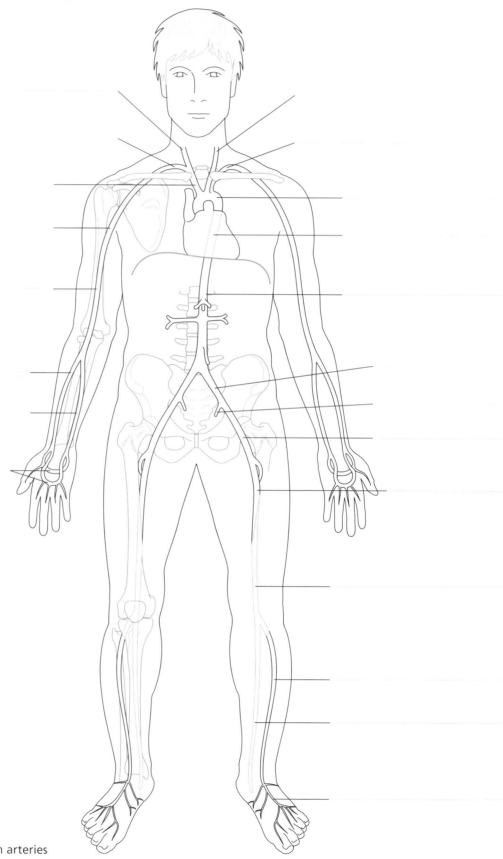

Figure 5.15 Aorta and main arteries

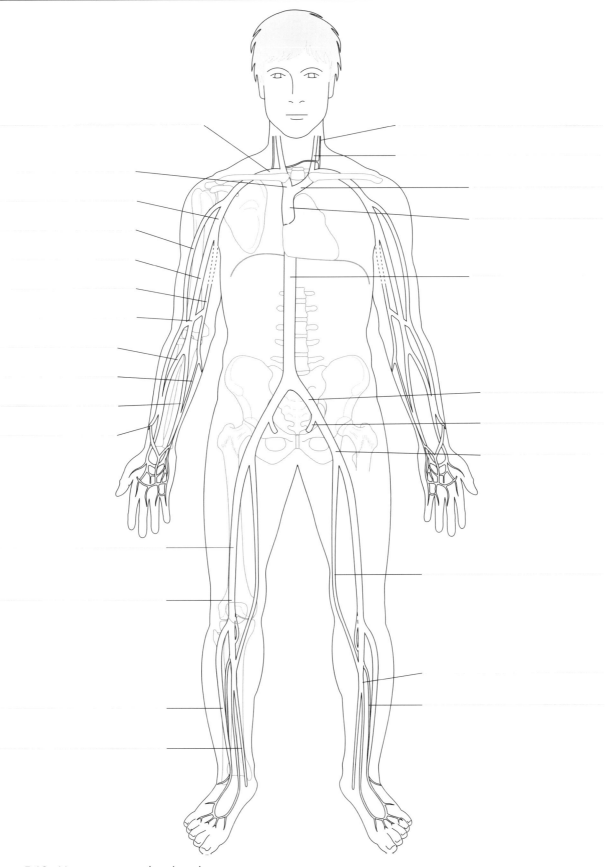

Figure 5.16 Venae cavae and main veins

Matching and labelling

73. Match and label the main arteries of the right arm using the key choices below (Figure 5.17).

Key choices:

Digital arteries
Vertebral artery
Right subclavian artery
Left subclavian artery
Deep palmar arch
Axillary artery
Superficial palmar arch
Left subclavian artery
Aorta
Internal thoracic artery
Common carotid artery
Brachial artery
Ulnar artery
Brachiocephalic artery

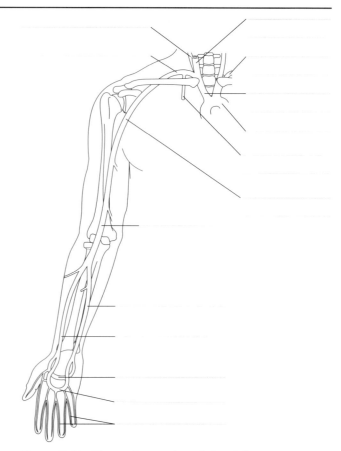

Figure 5.17 The main arteries of the right arm

74. Trace the flow of blood from the heart through the leg by putting the vessels listed below in the correct order, starting with the aorta and finishing with the inferior vena cava.

Femoral vein
Anterior tibial vein
Dorsalis pedis artery
Femoral artery
Digital arteries
External iliac vein
Common iliac artery
Popliteal vein
Dorsal venous arch
Popliteal artery
Common iliac vein
Anterior tibial artery

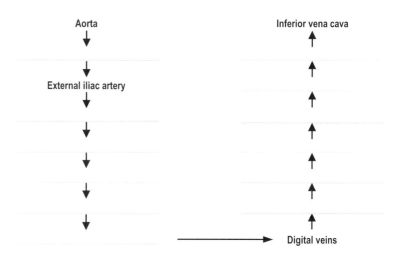

75. For each of the following major blood vessels named in Table 5.5, decide whether it is paired (P) or unpaired (U), and match with the correct statement from the box.

Artery	P or U	Correct statement
Internal iliac artery		
External iliac artery		
Intercostal artery		
Superior mesenteric artery		
Phrenic artery		
Coeliac artery		
Abdominal aorta		
Cystic artery		
Hepatic portal vein		

Table 5.5 Features of some major vessels

Statements:

Supplies the diaphragm

Becomes the femoral artery

Supplies the gall bladder

Splits into the right and left common iliac arteries

Divides into the hepatic, splenic and gastric arteries

Supplies the small intestine

Runs along the inferior border of each rib

Links the small intestine and the liver

Branches to supply the pelvic organs

  Matching, colouring and labelling

76. On Figure 5.18, label the structures indicated using the labels provided.

Abdominal aorta	Superior vena cava	Liver
Inferior vena cava	Pulmonary artery	Right atrium
Right ventricle	Umbilical vein	Ductus venosus
Foramen ovale	Lung	Umbilical cord
Hepatic portal vein	Ductus arteriosus	Pulmonary veins
Umbilicus	Aortic arch	Umbilical arteries
Placenta	Common iliac artery	

77. On Figure 5.18, insert arrows on the umbilical arteries and vein, pulmonary arteries and veins, right and left sides of the heart, aorta, foramen ovale and ductus arteriosus to show direction of blood flow through the fetal circulation.

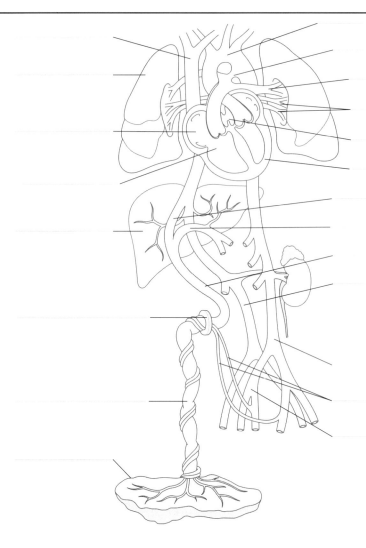

Figure 5.18 The fetal circulation

? Pot luck

78. List the functions of the placenta.

- _____

- _____

- _____.

The lymphatic system

The lymphatic system consists of a network of lymphatic vessels, the fluid that flows through them and various specialized organs and tissues. Its main functions are in tissue drainage and in the production and maintenance of immune cells.

   Matching, colouring and labelling

1. Figure 6.1 shows the main structures of the lymphatic system. Label the structures indicated using the key choices listed.

   Key choices:
   Inguinal nodes
   Lymphatic vessels
   Thymus gland
   Red bone marrow
   Palatine tonsil
   Thoracic duct (twice)
   Spleen
   Aggregated lymph follicles (Peyer's patches)
   Submandibular nodes
   Cisterna chyli
   Intestinal nodes
   Axillary nodes
   Right lymphatic duct

2. On Figure 6.1, colour the spleen, red bone marrow of the right femur and the thymus gland different colours, using the key below.

   ○ Spleen
   ○ Thymus gland
   ○ Red bone marrow

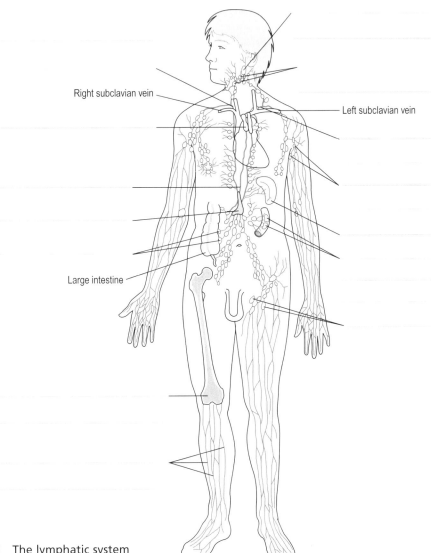

Right subclavian vein

Left subclavian vein

Large intestine

Figure 6.1 The lymphatic system

3. Outline the difference between interstitial fluid and lymph:

LYMPH

 MCQs

4. Which important constituent of plasma is absent from lymph? _____.

    a. Glucose   **b.** Plasma proteins   **c.** Hormones   **d.** Antibodies.

5. Which of the following is/are functions of the lymphatic system? (Choose all that apply) _____.

    a. Filtering of blood through lymph nodes
    b. Absorption of fat-soluble vitamins in the small intestine
    c. Maturation of lymphocytes
    d. Production of phagocytic macrophages.

6. Which nutrient is absorbed into the lymphatic vessels of the small intestine? _____.

    a. Glucose   **b.** Amino acids   **c.** Vitamins   **d.** Fats.

7. Lymphocytes, which circulate in the lymph, are produced in the: _____.

    a. Spleen   **b.** Thymus gland   **c.** Red bone marrow   **d.** Cisterna chyli.

LYMPHATIC VESSELS

 Completion

8. The following paragraph describes lymphatic vessels. Complete it by scoring out the incorrect options in bold, leaving the correct option(s).

The smallest lymphatic vessels are called **ducts/venules/capillaries**. One significant difference between them and the smallest blood vessels is that they **are only one cell thick/have permeable walls/originate in the tissues**; their function is to drain the lymph, containing **red blood cells/white blood cells/platelets**, away from the interstitial spaces. Most tissues have a network of these tiny vessels, but one notable exception is **bone tissue/ muscle tissue/fatty tissue**. The individual tiny vessels join up to form larger ones, which now contain **two/three/ four** layers of tissue in their walls, similar to veins in the cardiovascular system. The inner lining, the **endothelial/fibrous/muscular** layer, covers the valves, which **filter the lymph/store the lymph/regulate flow of lymph**. Unlike the cardiovascular system, there is no organ acting as a pump to push lymph through the vessels, but forward pressure is applied to the lymph by various mechanisms, including **movement of the valves pushing lymph onward/squeezing of the vessels by external structures like skeletal muscle/intrinsic contractility of the smooth muscle of lymphatic vessel walls**. As vessels progressively unite and become wider and wider, eventually they empty into the biggest lymph vessels of all, the **thoracic duct and the right lymphatic duct/subclavian duct and the right lymphatic duct/thoracic duct and subclavian duct**. The first one of these drains the **left side of the body/right side of the body above the diaphragm/lower limbs and pelvic area**. The second drains the **upper body above the pelvis/right side of the body/lower part of the body and the upper left side above the diaphragm**.

LYMPHATIC ORGANS AND TISSUES

  Labelling and colouring

9. Figure 6.2 shows the internal structure of a lymph node. Label the structures indicated and colour the capsule and associated trabeculae.

10. On Figure 6.2, insert arrows to show which way lymph will flow through this lymph node.

11. List four types of particulate matter filtered out by lymph nodes.

_____

_____

_____

_____.

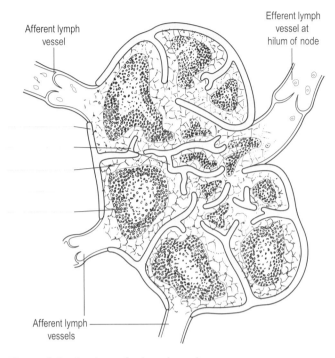

Figure 6.2 Section of a lymph node

 MCQs

12. The tissue in lymph nodes responsible for filtration is called: _____.

   a. Trabecular tissue
   b. Capsular tissue
   c. Reticular tissue
   d. Hilar tissue.

13. In the lymph node, lymphocytes: _____.

   a. Filter lymph
   b. Multiply
   c. Phagocytose bacteria
   d. Differentiate into monocytes and macrophages.

14. Which one of the following statements is NOT true concerning the structure of a lymph node? _____.

   a. They have no blood supply but receive their nutrients and oxygen from the lymph that passes through
   b. Lymph flows into the node through afferent lymph vessels
   c. Lymph nodes vary in size, the largest being about 25 mm across
   d. Each node has a concave surface called the hilum where various vessels enter and leave the node.

15. Which is the main group of nodes draining the head and neck? _____.

   a. The mammary nodes
   b. The axillary nodes
   c. The inguinal nodes
   d. The cervical nodes.

 Definition

16. Define the term phagocytosis: _____

_____.

Matching

17. For each of the following key choices, decide whether it applies to the lymph nodes, the spleen or the thymus, and complete Table 6.1.

Key choices:

T-lymphocytes mature here	Lies immediately below the diaphragm
Bean-shaped	At its maximum size in puberty
Largest lymphatic organ	Filters lymph
Site of multiplication of activated lymphocytes	Oval in shape
Stores blood	Lies immediately behind the sternum
Made up of two narrow lobes	Size from pinhead to almond size
Red blood cells destroyed here	Secretes the hormone thymosin
Distributed throughout lymphatic system	Synthesizes red blood cells in the fetus
Maximum weight usually 30–40 g	Phagocytoses cellular debris

Spleen	Thymus	Lymph node

Table 6.1 Characteristics of lymph nodes, spleen and thymus

? **MCQs**

18. Characteristically, mucosa-associated lymphoid tissue (MALT) is located: _____.

   a. in close association with groups of lymph nodes
   b. only in the gastrointestinal tract
   c. associated with organs exposed to the external environment
   d. in the central nervous system.

19. Tonsils: _____.

   a. contain B- and T-lymphocytes
   b. filter lymph
   c. are enclosed within a tonsillar capsule
   d. are supplied with no more than two afferent lymphatic vessels.

20. Aggregated lymphoid follicles (Peyer's patches) protect against: _____.

   a. inhaled bacteria
   b. blood-borne bacteria
   c. swallowed bacteria
   d. sexually transmitted bacteria.

21. Thymosin levels: _____.

   a. remain steady throughout life
   b. decline after adolescence
   c. increase with age
   d. peak twice: once in youth, and again in middle age.

The nervous system

The nervous system detects and quickly responds to changes inside and outside the body. Together with the endocrine system, it controls important aspects of body function. Responses to changes in the internal environment maintain homeostasis and regulate our involuntary functions. Responses to changes in the external environment maintain posture and other voluntary activities.

The nervous system consists of the brain, spinal cord and peripheral nerves organized in a way that enables rapid communication between different parts of the body. This chapter is designed to help you learn about the structure and functions of the nervous system and its components.

 Matching and labelling

1. Name the two main parts of the central nervous system:

   • _____   • _____

2. Label the sensory and motor neurones on Figure 7.1.

3. Insert the key choices beside the bullet points on Figure 7.1 showing their relationships to the nervous system.

Key choices:
Chemoreceptors
Glands
Sight
Hearing
Smooth muscle
Baroreceptors
Taste
Cardiac muscle
Skeletal muscle
Smell
Osmoreceptors

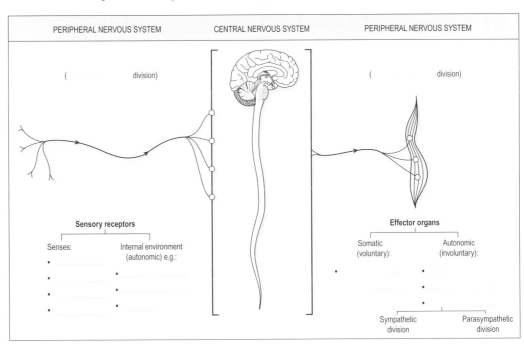

Figure 7.1 Functional components of the nervous system

73

NEURONES

 Labelling

4. Label the parts of the neurone indicated in Figure 7.2.

5. Draw an arrow beside a neurone in Figure 7.2 to show the direction of impulse conduction.

6. Outline the main difference between the structure of myelinated and non-myelinated neurones.

_____

_____

_____

_____

_____

_____ .

Myelinated neurone **Non-myelinated neurone**

Figure 7.2 The structure of neurones

 MCQs

7. The grey matter of the nervous system is formed by: _____.

   a. Cell bodies   **b.** Axons   **c.** Terminal boutons   **d.** Glial cells.

8. Groups of cell bodies in the central nervous system are usually described as: _____.

   a. Tracts   **b.** Nerves   **c.** Ganglia   **d.** Nuclei.

9. The white matter of the nervous system is formed by: _____.

   a. Cell bodies   **b.** Axons   **c.** Terminal boutons   **d.** Glial cells.

10. Nerve fibres that carry impulses from the central nervous system are described as: _____.

    a. Efferent   **b.** Afferent   **c.** Sensory   **d.** Tracts.

11. Groups of cell bodies in the peripheral nervous system are described as: _____.

    a. Tracts   **b.** Nerves   **c.** Ganglia   **d.** Nuclei.

12. The neurone receives incoming impulses at the: _____.

    a. Axon   **b.** Axon hillock   **c.** Dendrites   **d.** Terminal boutons.

Completion

13. Fill in the blanks in the paragraph below to describe the events that occur during conduction of nerve impulses.

Transmission of the _____, or impulse, is due to movement of _____ across the nerve cell membrane. In the resting state the nerve cell membrane is _____ due to differences in the concentrations of ions across the plasma membrane. This means that there is a different electrical charge on each side of the membrane, which is called the resting _____. At rest the charge outside the cell is _____ and inside it is _____. The principal ions involved are _____ and _____. In the resting state there is a continual tendency for these ions to diffuse down their _____. During the action potential, sodium ions flood _____ the neurone causing _____. This is followed by _____ when potassium ions move _____ the neurone. In myelinated neurones the insulating properties of the_____ prevent the movement of ions across the membrane where this is present. In these neurones, impulses pass from one _____ to the next and transmission is called _____. In unmyelinated fibres, nerve impulses are conducted by the process called _____. Impulse conduction is faster when the mechanism of transmission is _____ than when it is_____. The diameter of the neurone also affects the rate of impulse conduction: the _____ the diameter, the faster the conduction.

Colouring, matching and labelling

14. Colour and match the following on each part of Figure 7.3:

○ Presynaptic neurone
○ Postsynaptic neurone

15. Label the structures indicated on Figure 7.3.

16. Add arrows showing the direction of impulse transmission in the neurones shown on the main part of Figure 7.3.

17. Name the neurotransmitter at the neuromuscular

junction. _____.

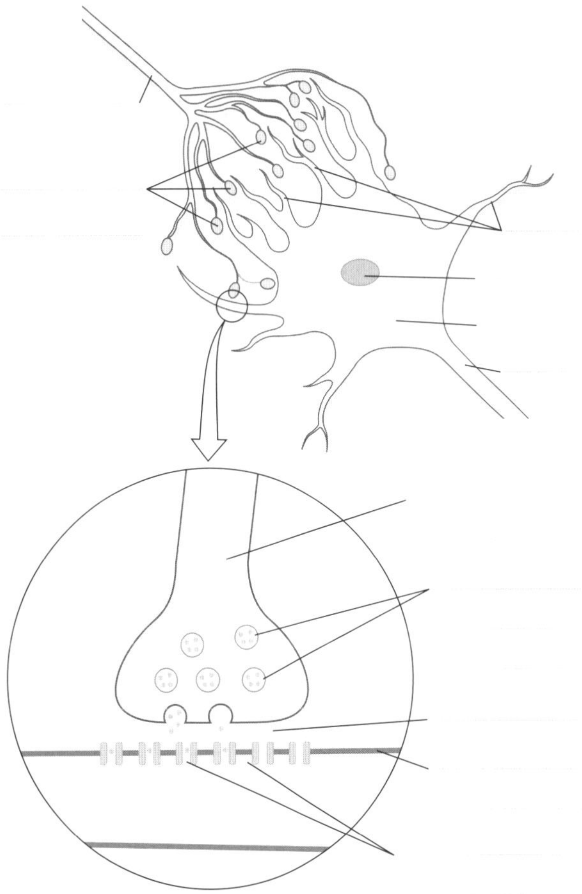

Figure 7.3 Diagram of a synapse

Completion

18. Fill in the blanks to complete the paragraph below, describing the transmission of an impulse from one neurone to the next.

The region where a nerve impulse passes from one neurone to another is called the _____. The distal end

of the presynaptic neurone breaks up into minute branches known as _____. These are in close

proximity to the dendrites and cell bodies of the _____. The space between them is the

_____. In the ends of the presynaptic neurones are spherical structures called _____

containing chemicals known as the _____. When the action potential depolarizes the presynaptic

membrane, the chemicals in the membrane-bound packages are released into the synaptic cleft by the process of

_____. The chemicals released then move across the synaptic cleft by _____. They act on

specific areas of the postsynaptic membrane called _____ causing _____.

  Colouring and matching

19. Colour and match the items below with the corresponding structures in Figure 7.4.

○ Axon
○ Nerve
○ Blood vessels
○ Perineurium
○ Epineurium
○ Endoneurium

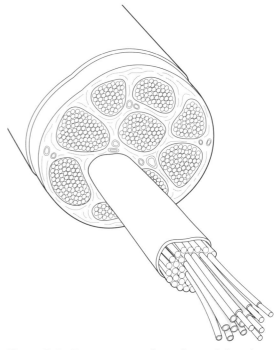

Figure 7.4 Transverse section of a peripheral nerve showing the protective coverings

CENTRAL NERVOUS SYSTEM

 Completion

20. Complete Table 7.1 by ticking the appropriate column(s) for each statement about the meninges.

	Dura mater	Arachnoid mater	Pia mater
Consists of two layers of dense fibrous tissue			
Consists of fine connective tissue			
The epidural space lies above this layer			
The subdural space lies between these two layers			
Surrounds the venous sinuses			
The subarachnoid space separates these two layers			
Forms the filum terminale			
CSF is found in the space between these two layers			
Equivalent to the periosteum of other bones			

Table 7.1 Characteristics of the meninges

 Labelling

21. Label the meninges and other structures indicated on Figure 7.5.

22. Outline the function of the blood–brain barrier.

_____

_____

_____.

23. What is access to the epidural space used for in clinical medicine?

_____

_____

_____.

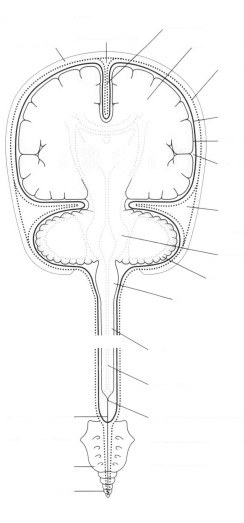

Figure 7.5 The meninges covering the brain and spinal cord

Colouring and labelling

24. Colour the ventricular system of the brain.

25. Label the components of the ventricular system identified in Figure 7.6.

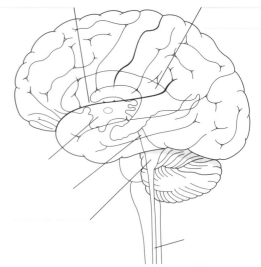

Figure 7.6 The positions of the ventricles in the brain viewed from the left side

26. Name the fluid found in the ventricles of the brain _____.

? **MCQs**

27. CSF is secreted by choroid plexuses situated in: _____.

   a. The ventricles   **b.** The dura mater   **c.** The pia mater   **d.** The arachnoid mater.

28. CSF circulation is assisted by (choose all that apply) _____.

   a. Breathing   **b.** Pulsation of blood vessels   **c.** A pump   **d.** Changes in position.

29. CSF returns to the blood: _____.

   a. through the foramina in the roof of the fourth ventricle when venous pressure is greater than CSF pressure
   b. through the foramina in the roof of the fourth ventricle when CSF pressure is greater than venous pressure
   c. through the arachnoid villi when venous pressure is greater than CSF pressure
   d. through the arachnoid villi when CSF pressure is greater than venous pressure.

30. The approximate volume of CSF in adults is: _____.

   a. 50 ml   **b.** 100 ml   **c.** 150 ml   **d.** 200 ml.

31. Normal constituents of CSF include: _____.

   a. Glucose, erythrocytes, leukocytes, albumin   **c.** Glucose, bilirubin, albumin, mineral salt
   b. Mineral salts, erythrocytes, leukocytes, water   **d.** Water, glucose, albumin, leukocytes.

32. Normal CSF pressure when lying down is around: _____.

   a. 5 cm H_2O **b.** 10 cm H_2O **c.** 15 cm H_2O **d.** 20 cm H_2O.

? **Pot luck**

33. Outline the functions of CSF:

_____

_____

_____

BRAIN

   Colouring, matching and labelling

34. Colour and match the following parts of the brain on Figure 7.7.

- ○ Cerebrum
- ○ Diencephalon
- ○ Brain stem
- ○ Cerebellum

35. Label the other structures shown on Figure 7.7.

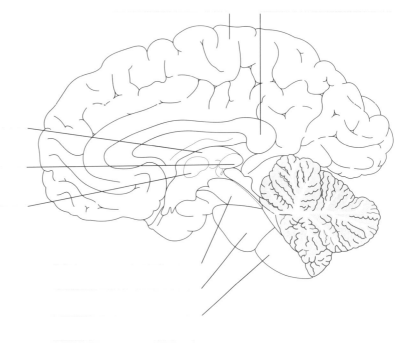

Figure 7.7 The parts of the central nervous system

 Pot luck

36. Name the structure which enables the blood supply to the brain to be maintained even when one of the

supplying arteries is blocked _____.

 Completion

37. Complete the blanks to describe the structure of the cerebrum.

This is the largest part of the brain and is divided into left and right _____. Deep inside, the

two parts are connected by the _____, which consists of _____ matter.

The superficial layer of the cerebrum is known as the _____ and consists of

nerve _____ or _____ matter. The deeper layer consists of nerve _____ and is _____ in

colour. The cerebral cortex has many furrows and folds that vary in depth. The exposed areas are the convolutions

or _____ and they are separated by _____, also known as _____. These convolutions

increase the _____ of the cerebrum.

Colouring and matching

38. Colour the structures listed below, matching them with those on Figure 7.8.

- ○ Corpus callosum
- ○ Basal ganglia
- ○ Thalamus
- ○ Internal capsule
- ○ Cerebral cortex
- ○ Hypothalamus

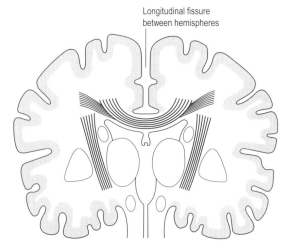

Longitudinal fissure between hemispheres

Figure 7.8 A section of the cerebrum showing some connecting nerve fibres

39. Outline the main functions of the cerebrum:

_____ .

_____ .

   ## Colouring, matching and labelling

40. Colour, match and label the functional areas of the cerebrum listed below with those identified in Figure 7.9.

- ○ Taste area
- ○ Somatosensory area
- ○ Primary motor area
- ○ Frontal area
- ○ Sensory speech (Wernicke's) area
- ○ Auditory area
- ○ Motor speech (Broca's) area
- ○ Premotor area
- ○ Visual area

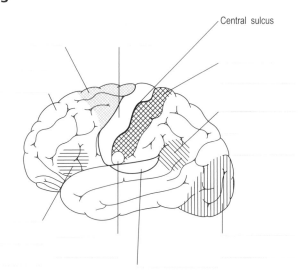

Central sulcus

Figure 7.9 The cerebrum showing the functional areas

 ## Pot luck

41. Which parts of neurones form:

a. The grey matter of the cerebral cortex? _____

b. The white matter of the cerebral cortex? _____ .

Completion

42. The following paragraphs describe aspects of the motor areas of the cerebrum. Cross out the wrong options so that it reads correctly.

The primary motor area lies in the **parietal/temporal/frontal** lobe immediately anterior to the **central/lateral/ parieto-occipital** sulcus. The cell bodies are **oval/pyramid-shaped/hexagonal** and stimulation leads to contraction of **smooth/skeletal/cardiac** muscle. Their nerve fibres pass downwards through the **thalamus/internal capsule/ hypothalamus** to the **midbrain/cerebellum/medulla** where they cross to the opposite side then descend in the spinal cord. These neurones are the upper motor neurones. They synapse with the lower motor neurones in the **spinal cord/medulla/cerebellum** and lower motor neurones terminate at a **neuromuscular junction/ synapse/ sensory receptor**. This means that the motor area of the right hemisphere controls skeletal muscle movement on **the left side/the right side/both sides** of the body.

In the motor area of the cerebrum, body areas are represented **in mirror image/the right way up/upside down** and the proportion of the cerebral cortex that represents a particular part of the body reflects its **size/complexity of movement/distance from the brain**.

Broca's area lies in the **parietal/temporal/frontal** lobe and controls the movements needed for **swallowing/ writing/speech**. The right hemisphere is dominant in **left-handed/ambidextrous/right-handed** people.

The frontal area is situated in the **parietal/frontal/temporal** lobe and is thought to be involved in one's **body clock/feelings of hunger/character**.

 Matching

43. Match the statements below with the sensory areas of the cerebrum listed (you will need some sensory areas more than once).

Olfactory area
Gustatory area
Visual area
Auditory area
Sensory speech area

a. The right side is dominant in left-handed people:

_____.

b. The centre for perception of taste:

_____.

c. Receives impulses from the cochlear nerves:

_____.

d. The centre for perception of smell:

_____.

e. Sensory receptors are located in the retina:

_____.

f. The centre for sight:

_____.

g. Sensory receptors are located in the inner ear:

_____.

h. Sensory nerves are activated by dissolved chemicals:

_____.

i. The most posterior sensory area of the brain:

_____.

j. Situated in the temporal lobe:

_____.

? **MCQs**

44. Which of the following are part of the diencephalon (choose all that apply): _____.

    a. Midbrain   **b.** Pineal gland   **c.** Hypothalamus   **d.** Thalamus.

45. Which part of the diencephalon forms the lateral walls of the third ventricle? _____.

    a. Midbrain   **b.** Pineal gland   **c.** Hypothalamus   **d.** Thalamus.

46. Which area of the brain controls the pituitary gland? _____.

    a. Hypothalamus   **b.** Thalamus   **c.** Medulla oblongata   **d.** Pons.

47. The functions of the pons include (choose all that apply): _____.

    a. Secretion of CSF   **c.** Contributes to control of water and electrolyte balance

    b. Relay station for nerve impulses   **d.** Contributes to control of breathing.

48. The cardiovascular centre is located in the _____.

    a. Medulla oblongata   **b.** Pons   **c.** Hypothalamus   **d.** Heart.

49. The vital centres include (choose all that apply): _____.

    a. Vomiting centre   **b.** Vasomotor centre   **c.** The satiety centre   **d.** Temperature regulating centre.

50. The functions of the hypothalamus include control of (choose all that apply): _____.

    a. Secretion of hormones from the posterior pituitary gland   **c.** Body temperature
    b. Secretion of hormones from the anterior pituitary gland   **d.** Appetite.

51. Proprioceptor impulses originate in the (choose all that apply): _____.

    a. Muscle   **b.** Skin   **c.** Joints   **d.** Eye.

SPINAL CORD

? **Pot luck**

52. Outline the functions of:

    a. The reticular activating system _____

_____.

    b. The cerebellum _____

_____.

53. What is the length of the spinal cord _____.

54. The distal end of the spinal cord lies at the lower border of the _____.

55. What is a lumbar puncture? _____

_____.

56. Identify the points of origin and destination of the following tracts:

    a. Spinothalamic: origin_____; destination _____.

    b. Corticospinal: origin _____; destination _____.

Completion

57. Tick the appropriate boxes in Table 7.2 to indicate whether each statement relates to either sensory or motor pathways that travel through the spinal cord.

	Motor pathways	Sensory pathways
Impulses travel towards the brain		
The extrapyramidal tracts are an example of these		
Consist of two neurones		
Contain afferent tracts		
Their fibres pass through the internal capsule		
Impulses from proprioceptors travel via these pathways		
Are involved in fine movements		
Are involved in movement of skeletal muscles		
Equivalent to the periosteum of other bones		
Impulses follow activation of receptors in the skin		
Impulses travel away from the brain		
May consist of either two or three neurones		

Table 7.2 Characteristics of the motor and sensory pathways of the spinal cord

Completion and labelling

58. Label the structures identified on Figure 7.10.

59. Draw arrows indicating the direction of the nerve impulses in a reflex arc on Figure 7.10.

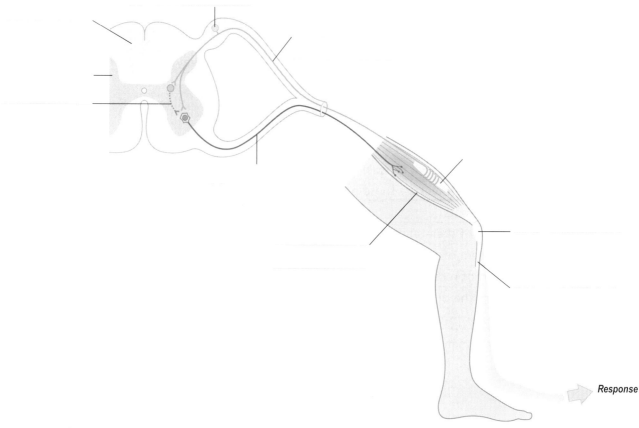

Response

Figure 7.10 The knee jerk reflex

60. Draw an arrow on Figure 7.10 showing where the knee jerk reflex is tested.

61. Which component shown on Figure 7.10 is absent in the stretch (knee jerk) reflex? _____.

PERIPHERAL NERVOUS SYSTEM

 Completion

62. Complete the following paragraph, describing the peripheral nervous system, by filling in the blanks.

Within the peripheral nervous system there are _____ pairs of spinal nerves and _____ pairs of cranial nerves. These nerves are composed of _____ nerve fibres conveying afferent impulses to _____ from _____ organs, or _____ nerve fibres that transmit efferent impulses from _____ to _____ organs. Some nerves, known as _____ nerves, contain both types of fibres.

63. Explain the function of a nerve plexus.

 Labelling

64. Name the plexuses and other structures shown on Figure 7.11.

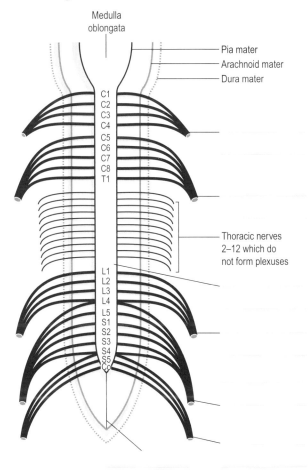

Figure 7.11 The meninges covering the spinal cord, spinal nerves and the plexuses they form

 **Labelling and matching**

65. Label and match the following nerves of the arm shown on the anterior view of Figure 7.12:

Ulnar	Radial	Median

66. Label and match the following nerves of the arm shown on the posterior view of Figure 7.12:

Ulnar (× 2 labels)	Radial (× 2 labels)	Axillary

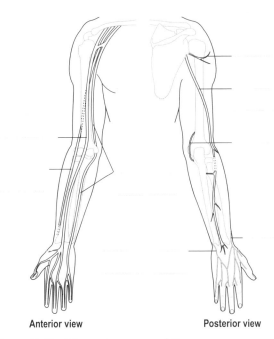

Anterior view   Posterior view

Figure 7.12 The main nerves of the arm

Labelling and matching

67. Label and match the following nerves of the leg shown on the anterior view of Figure 7.13:

Sural
Femoral
Obturator
Common peroneal
Saphenous
Deep peroneal
Superficial peroneal
Lateral cutaneous nerve of thigh

68. Label and match the following nerves of the leg shown on the posterior view of Figure 7.13:

Common peroneal
Posterior cutaneous nerve of thigh
Sural
Sciatic
Tibial (× 2 labels)

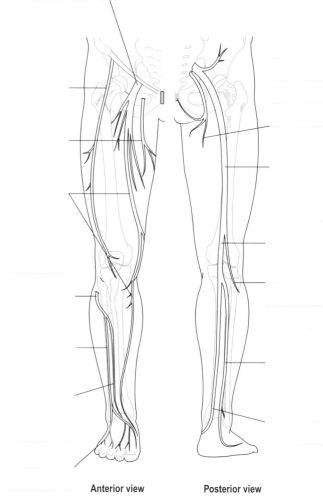

Anterior view   Posterior view

Figure 7.13   The main nerves of the leg

 Completion

69. Complete Table 7.3 by inserting the names of the nerves supplying the muscles in column 1 and the plexuses from which they arise.

Muscle supplied	Name of nerve	Plexus of origin
Intercostal muscles		N/A
Diaphragm		
Quadriceps		
Hamstrings		
External anal sphincter		
External urethral sphincter		

Table 7.3   Nerves supply to muscles

Colouring and labelling

70. Identify the parts of the central nervous system indicated on the left side of Figure 7.14.

71. Colour the cranial nerves and their associated structures on Figure 7.14.

72. Name the numbered cranial nerves on the right side of Figure 7.14.

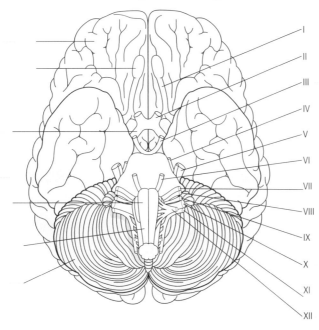

I
II
III
IV
V
VI
VII
VIII
IX
X
XI
XII

Figure 7.14 The inferior surface of the brain showing the cranial nerves

 Completion

73. Insert the names and functions of the cranial nerves in the appropriate boxes in Table 7.4.

74. Identify the type of each cranial nerve (sensory, motor or mixed) to complete Table 7.4.

Number	Name	Function	Type
I			
II			
III			
IV			
V			
VI			
VII			
VIII			
IX			
X			
XI			
XII			

Table 7.4 The cranial nerves and their functions

? MCQs

75. Which cranial nerve innervates most of the gastrointestinal tract? _____.

    a. Trochlear   **b.** Abdjucent   **c.** Vagus   **d.** Accessory.

76. Which cranial nerves carry impulses from the receptors for smell? _____.

    a. Optic   **b.** Occulomotor   **c.** Glossopharyngeal   **d.** Olfactory.

77. Which cranial nerves are necessary for non-verbal gestures? _____.

    a. Facial   **b.** Trigeminal   **c.** Vagus   **d.** Olfactory.

78. Which cranial nerves innervate most of the muscles of the eye? _____.

    a. Optic   **b.** Occulomotor   **c.** Ophthalmic   **d.** Facial.

79. Which cranial nerves convey the pain from toothache? _____.

    a. Hypoglossal   **b.** Trigeminal   **c.** Abducent   **d.** Facial.

80. Which cranial nerves convey the pain from grit in the eye? _____.

    a. Optic   **b.** Occulomotor   **c.** Trigeminal   **d.** Abducent.

AUTONOMIC NERVOUS SYSTEM

? Pot luck

81. Name the effector organs of the autonomic nervous system:

    • _____   • _____   • _____.

82. List the two divisions of the autonomic nervous system:

    • _____   • _____.

83. Decide whether each of the following statements is TRUE (T) or FALSE (F):

    a. The sympathetic nervous system is sometimes referred to as the craniosacral outflow: _____

    b. The parasympathetic nervous system is associated with fight or flight responses: _____

    c. The sympathetic nervous system has a preganglionic and a postganglionic neurone: _____

    d. The neurotransmitter at the sympathetic ganglia is noradrenaline (norepinephrine) _____

    e. The neurotransmitter at the parasympathetic ganglionic synapse is acetylcholine: _____

    f. Stimulation of the parasympathetic nervous system results in release of noradrenaline (norepinephrine) from the adrenal glands: _____

    g. There is no sympathetic nerve supply to sweat glands: _____

    h. In the parasympathetic nervous system, the preganglionic fibre is longer than the postganglionic fibre: _____

    i. The prevertebral ganglia are part of the parasympathetic nervous system: _____

    j. The parasympathetic neurotransmitter at effector organs is acetylcholine: _____

    k. The autonomic nervous system is involved in voluntary functions: _____.

Colouring and labelling

84. Draw in lines to represent the postganglionic sympathetic fibres on Figure 7.15.

85. Label the three prevertebral ganglia shown on Figure 7.15.

86. Colour and name the structures supplied by the sympathetic nervous system shown in Figure 7.15.

87. Complete Figure 7.15 by inserting the effects of sympathetic stimulation in the right-hand column.

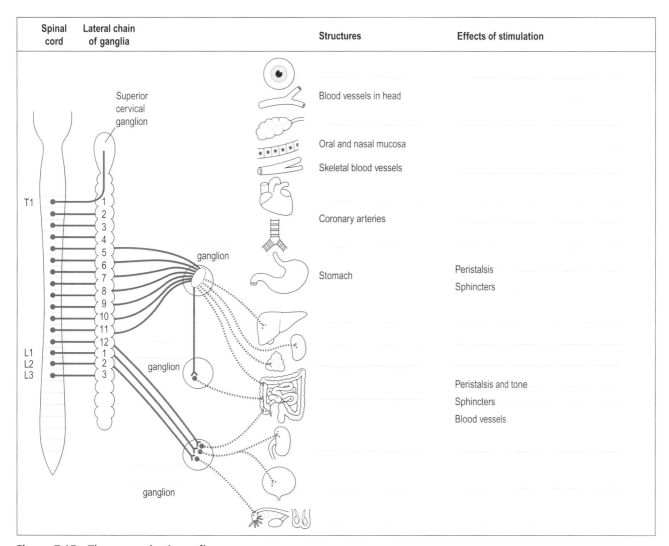

Figure 7.15 The sympathetic outflow

 ## Pot luck

88. What is referred pain?

_____

89. Explain when this occurs.

_____

Colouring and labelling

90. Draw in lines representing the postganglionic fibres on Figure 7.16.

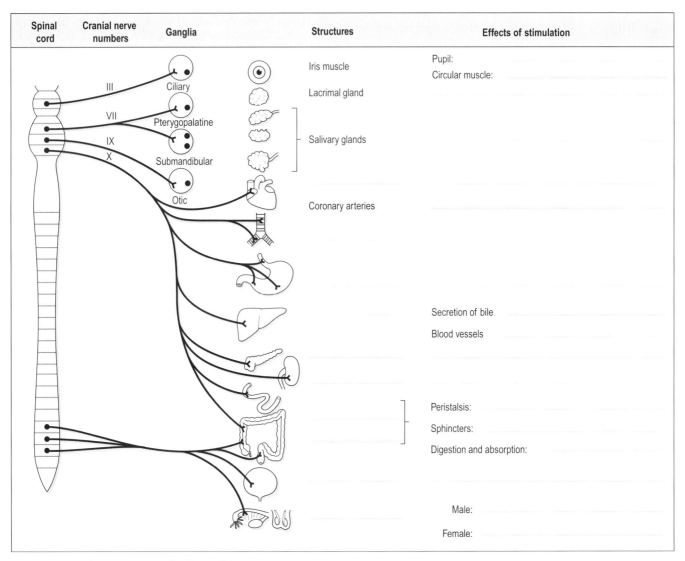

Figure 7.16 The parasympathetic outflow

91. Explain why only some of the organs on Figure 7.16 appear to have postganglionic parasympathetic fibres.

92. Colour and label the structures innervated by the parasympathetic nervous system.

93. Complete Figure 7.16 by inserting the effects of parasympathetic stimulation in the right-hand column.

The special senses

The special senses are those of hearing, balance, sight, smell and taste. For each one, there are specialized sensory receptors located within sensory organs in the head. Incoming information is transmitted to the brain and together with information from other parts of the brain, for example the memory, it is integrated and an effector response ensues. These senses often work together either consciously or subconsciously. Conscious effects include both the taste and smell of foods, which are usually, but not always, associated with enjoyment. At the same time they subconsciously prepare the digestive system for action. This chapter will help you learn about the special senses.

HEARING AND THE EAR

   Colouring, matching and labelling

1. Colour and match the following parts of the ear:

- ○ Outer ear
- ○ Middle ear
- ○ Inner ear

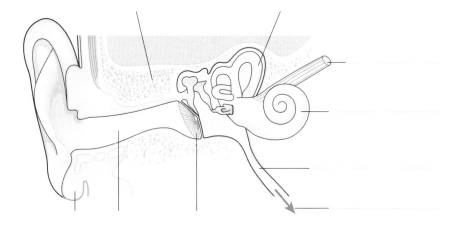

Figure 8.1 The parts of the ear

2. Label the structures shown on Figure 8.1.

3. Name the substance secreted by ceruminous glands in the auditory canal. _____.

4. Colour and match the following parts of the middle ear shown on Figure 8.2:

- ○ Tympanic cavity
- ○ Tympanic membrane
- ○ Oval window
- ○ Malleus (hammer)
- ○ Incus (anvil)
- ○ Stapes (stirrup)

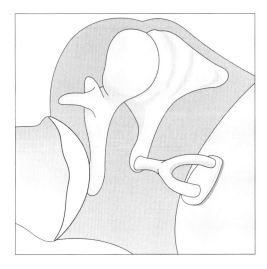

Figure 8.2 The middle ear

Colouring, labelling and matching

5. Colour and match the following parts on Figure 8.3:

○ Bony labyrinth	○ Membranous labyrinth	○ Temporal bone

6. Label the structures indicated on Figure 8.3.

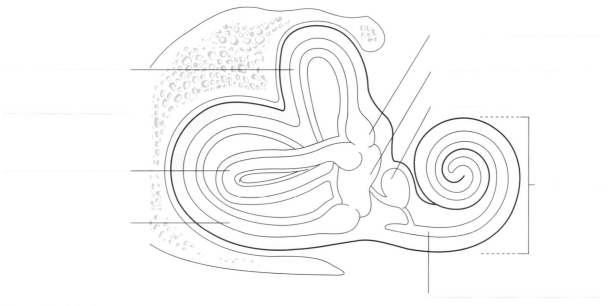

Figure 8.3 The inner ear

7. Which part of the bony labyrinth contains the oval and round windows in its lateral wall? _____.

8. On Figure 8.4 colour the part(s) of the inner ear
   containing:

   ○ Endolymph
   ○ Perilymph

9. Label the structures indicated on Figure 8.4.

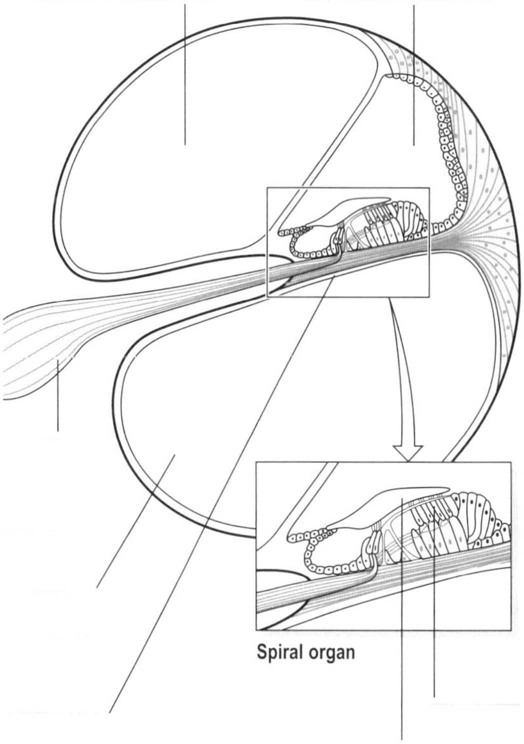

Spiral organ

Figure 8.4   A cross section of the cochlea
showing the spiral organ (of Corti)

? Pot luck

10. Decide whether each of the statements below is TRUE (T) or FALSE (F):

    a. Of the ossicles, the malleus is also known as the hammer _____.

    b. The ossicles are situated in the inner ear _____.

    c. The scala media is also known as the cochlear duct _____.

    d. The cochlea contains the utricle and saccule _____.

    e. The auditory receptors are axons of the sensory neurones that combine to form the cochlear nerve _____.

    f. Perception of sound in the cerebrum takes place only on the same side as the auditory receptors
       stimulated _____.

✎ Completion

11. Fill in the blanks to describe the physiology of hearing.

A sound produces _____ in the air. The auricle _____ and _____ them along the _____

to the _____. The vibrations are _____ and _____ through the middle ear

by movement of the _____. At its medial end, movement of the _____ in the _____ window

sets up fluid waves in the _____ of the scala vestibuli. Most of this pressure is transmitted into the

_____ resulting in a corresponding fluid wave in the _____. This stimulates the auditory

receptors in the _____ cells in the organ of hearing, the _____. Stimulation of the auditory

receptors results in the generation of _____ that travel to the brain along the _____ part

of the _____ nerve. The fluid wave is extinguished by vibration of the membrane of the

_____ window.

12. Sound has the properties of pitch and volume. Insert the units of measurement on the axes of Figure 8.5.

13. On Figure 8.5A, draw in a sound wave corresponding with a low pitch sound.

14. On Figure 8.5B, draw in a sound wave corresponding with a low volume sound.

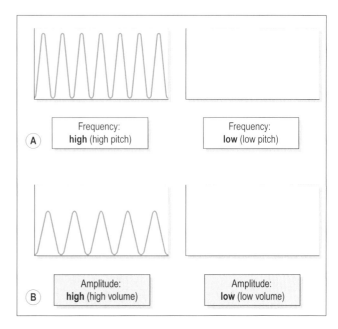

Figure 8.5 Behaviour of sound waves

BALANCE AND THE EAR

 Pot luck

15. There are *seven* errors in the paragraph below. Identify and correct them.

The organs involved with balance are found in the middle ear. They are the three round canals, one in each plane of space, and the vestibule which comprises two parts, the stapes and the utricle. The canals, like the cochlea, are composed of an outer bony wall and inner membranous ducts. The membranous ducts contain perilymph and are separated from the bony wall by endolymph. They have dilated portions near the vestibule called ampullae containing hair cells with sensory nerve endings between them. Any change in the position of the head causes movement in the endolymph and perilymph. This causes stimulation of the hair cells and nerve impulses are generated. These travel in the vestibular part of the vestibulocochlear nerve to the medulla via the cochlear nucleus. Perception of body position occurs because the cerebrum co-ordinates impulses from the eyes and proprioceptors in addition to those from the cerebellum.

SIGHT AND THE EYE

  Colouring, matching and labelling

16. Colour and match the following parts of the eye on Figure 8.6:

○ Retina	○ Optic nerve	○ Cornea
○ Sclera	○ Retinal blood vessels	○ Lens
○ Choroid	○ Anterior chamber	○ Vitreous body

17. Label the structures indicated on Figure 8.6.

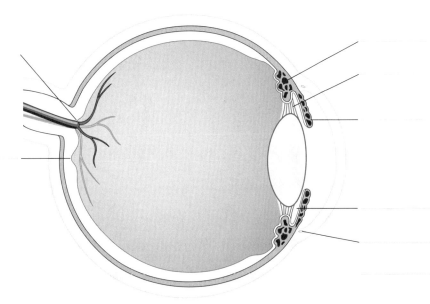

Figure 8.6   Section of the eye

  Colouring and matching

18. Colour and match the light sensitive receptors on Figure 8.7:

○ Rod shaped nerve cell
○ Cone shaped nerve cell

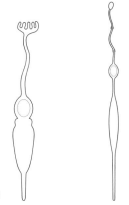

Figure 8.7   Rods and cones

 Completion

19. Fill in the blanks to describe the interior of the eye.

The anterior segment of the eye is incompletely divided into the _____ and _____ chambers by the _____. Both chambers contain _____ secreted into the _____ chamber by the _____. It circulates in front of the _____ and through the _____ into the _____ chamber and returns to the circulation through the _____. As there is continuous production and drainage, the intraocular pressure remains fairly constant. The structures in the front of the eye including the _____ and the _____ are supplied with nutrients by the _____. The posterior segment of the eye lies behind the _____ and contains the _____. It has the consistency of _____ and provides sufficient intraocular pressure to keep the eyeball from collapsing.

Completion

20. The retina lines the _____. Near the centre

is the _____ or yellow spot, consisting

only of _____-shaped cells. The small area of

the retina where the optic nerve leaves is the

_____ or _____.

 Labelling

21. Identify the parts of the optic pathways shown in Figure 8.8.

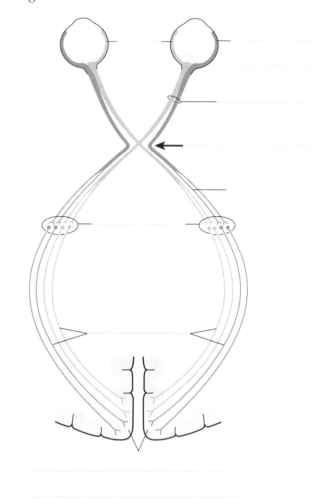

Figure 8.8 The optic nerves and their pathways

 MCQs

22. Light waves travel at: _____.

   a. 300 000 metres per second
   b. 300 000 metres per hour

   c. 300 000 kilometres per hour
   d. 300 000 kilometres per second.

23. Light waves of which colour have the longest wavelength? _____.

   a. Red
   b. Yellow
   c. Blue
   d. Violet.

24. Light enters the eyes by: _____.

   a. Refraction
   b. Accommodation
   c. Reflection
   d. Radiation.

25. Focusing of light is greatest when it passes through: _____.

   a. Plain glass
   b. A biconcave lens
   c. A biconvex lens
   d. Water.

26. Which of the following structures is able to change its refractory power? _____ .

    a. Vitreous body   **b.** Lens   **c.** Cornea   **d.** Conjunctiva.

27. Colour vision is discriminated by light sensitive pigments found in: _____ .

    a. Visual purple   **b.** Rods   **c.** Cones   **d.** Rods and cones.

28. An object appears white when: _____ .

    a. Light waves of all wavelengths are reflected   **c.** Microwaves are reflected
    b. Light waves of all wavelengths are absorbed   **d.** Gamma rays are absorbed.

29. Which structures in the eye have no blood supply? (choose all that apply) _____ .

    a. Retina   **b.** Iris   **c.** Lens   **d.** Cornea.

Matching and labelling

30. Insert the key choices into the appropriate spaces on the electromagnetic spectrum shown in Figure 8.9.

> *Key choices:*
> UV (ultraviolet) waves   Microwaves
> Radio waves   Gamma rays
> X-rays   Infrared rays

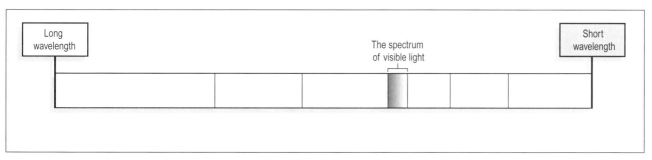

Figure 8.9 The electromagnetic spectrum

Colouring and matching

31. Colour and match the following parts of Figure 8.10A:

○ Ciliary body
○ Lens
○ Vitreous body

32. Label the two structures indicated on Figure 8.10A.

33. Complete Figure 8.10B by drawing in the changes that take place for near vision.

34. At what distance is focusing required for near vision? _____ .

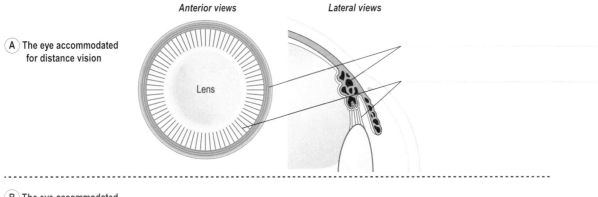

Anterior views   Lateral views

(A) The eye accommodated
for distance vision

Lens

(B) The eye accommodated
for near vision

Figure 8.10   The shape of the lens. **A.** Distant vision. **B.** Near vision

 Completion

35. Fill in the blanks to describe the factors that affect the size of the pupils.

The amount of light entering the eye is controlled by the _____ of the pupils. In a bright light they are

_____ and in darkness they are _____. The iris consists of two layers of smooth muscle –

contraction of the circular fibres causes _____ of the pupil while contraction of the radiating fibres causes

_____. The autonomic nervous system controls the size of the pupil – sympathetic stimulation causes

_____ while parasympathetic stimulation causes _____ of the pupil.

Pot luck

36. Name the three adjustments that the eyes make to focus on near objects (accommodation):

- _____

- _____

- _____ .

37. When moving from a very bright environment to a darkened one 'dark adaptation' occurs. Explain what this is and why it happens.

_____

_____ .

 Completion

38. Complete Table 8.1 by inserting the action of each of the extrinsic muscles of the eye.

Extrinsic muscle	Action
Medial rectus	
Lateral rectus	
Superior rectus	
Inferior rectus	
Superior oblique	
Inferior oblique	

Table 8.1   Actions of the extrinsic muscles of the eye

Colouring and matching

39. Colour and match the following on Figure 8.11:

- ○ Lacrimal gland
- ○ Upper and lower eyelids
- ○ Maxilla
- ○ Frontal bone
- ○ Optic nerve
- ○ Vitreous body
- ○ Lens
- ○ Tarsal plates

40. Emphasize the conjunctiva brightly.

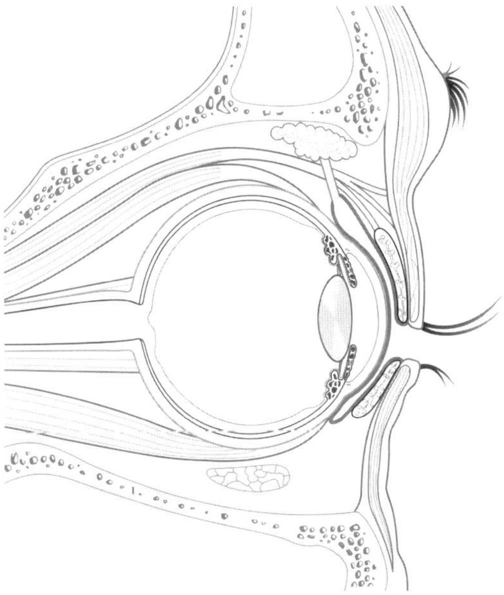

Figure 8.11   Section of the eye and its accessory structures

Pot luck

41. Which structures secrete tears? _____ .

42. List the constituents of tears: _____ .

43. State four functions of tears:

- _____

- _____

- _____

- _____

SENSE OF SMELL

 Pot luck

44. There are *five* errors in the paragraph below. Identify and correct them.

All odorous materials give off inert molecules that are carried into the nose in the inhaled air and stimulate the olfactory osmoreceptors. When currents of air are carried to the olfactory tract the smell receptors are stimulated, setting up impulses in the olfactory nerve endings. These pass through the cribriform plate of the mandible to the olfactory bulb. Nerve fibres that leave the olfactory bulb form the olfactory tract. This passes posteriorly to the olfactory lobe of the cerebellum where the impulses are interpreted and odour perceived.

Definitions

Define the following:

45. Anosmia _____ .

46. 'Adaption' to smell _____

_____ .

SENSE OF TASTE

 Completion

47. Fill in the blanks in the paragraph below describing the sense of taste.

Taste buds contain sensory receptors called _____. They are situated in the papillae of the

_____ and in the epithelia of the tongue, _____, _____ and _____. Some

of the taste buds have hair-like _____ on their free border projecting towards tiny pores in the

epithelium. Sensory receptors are stimulated by chemicals dissolved in _____ and _____

are generated when stimulation occurs. These are conducted to the brain where taste is perceived by the

_____ area in the _____ lobe of the cerebral cortex.

48. State how the sense of taste prepares the digestive system for action. _____.

49. Outline the protection function of taste. _____.

Colouring and matching

50. Colour and match the following parts of taste buds on Figure 8.12B:

○ Taste cells
○ Supporting cells
○ Nerve fibres
○ Epithelial cells of the tongue

51. Draw in the taste 'hairs'.

52. List four different tastes that we perceive:

- _____
- _____
- _____
- _____.

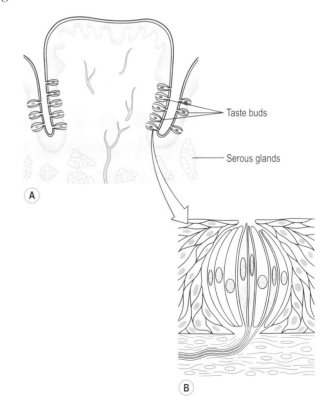

Figure 8.12 **Structure of taste buds. A.** A section of a papilla. **B.** A taste bud – greatly magnified

Definitions

Define the following terms:

53. Astigmatism

_____

_____

_____

_____ .

54. Myopia

_____

_____

_____ .

55. Hypermetropia

_____

_____

_____

_____ .

 Applying what you know

56. Draw in the lenses and altered light waves that will correct the refractive errors of the eye to complete Figures 8.13C and E.

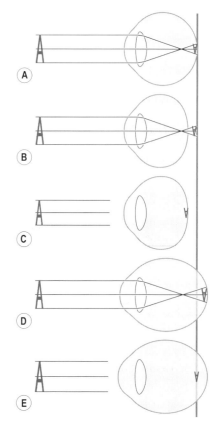

Figure 8.13
Common refractive errors of the eye and corrective lenses. **A** Normal eye. **B and C** Farsightedness. **D and E** Nearsightedness

The endocrine system

The endocrine system consists of ductless glands that secrete hormones. Together with the autonomic nervous system, the endocrine system maintains homeostasis of the internal environment and controls involuntary body functions. This chapter will help you explore the components of the endocrine system and their functions.

  Colouring and matching

1. Colour and match the endocrine glands identified on Figure 9.1:

   - ○ Adrenal glands
   - ○ Ovaries (in female)
   - ○ Pancreatic islets
   - ○ Pineal body
   - ○ Parathyroid glands
   - ○ Pituitary gland
   - ○ Testes (in male)
   - ○ Thymus gland
   - ○ Thyroid gland

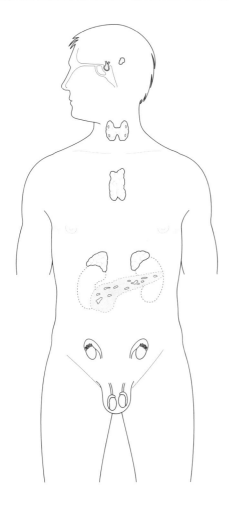

2. How many parathyroid glands are there?_____ .

3. Describe the location of the parathyroid glands:

   _____

   _____

 Completion

4. The paragraph below describes aspects of the endocrine system. Cross out the wrong options so that it reads correctly.

A hormone is formed by a gland that **secretes/excretes** it into **a duct/the bloodstream** which transports it to its target organ. When it arrives at its site of action, it binds to specific sites known as **enzymes/receptors** to bring about its effect. Many actions of the endocrine system are concerned with maintaining homeostasis of the **external/internal** environment. This often takes place in co-ordination with the **nervous/digestive** system. The effects of the endocrine system are usually **faster/slower** and **less/more** precise than the other system. Hormones may be either lipid-based, e.g. **adrenaline/steroids**, or peptides, which are **water/fat** soluble, e.g. **insulin/thyroid hormones**.

Figure 9.1 Positions of the endocrine glands

PITUITARY GLAND AND HYPOTHALAMUS

 Matching

5. Match the key choices to the statements in Table 9.1.

Key choices:		
Anterior lobe of the pituitary	Posterior lobe of the pituitary	Pituicyte
Intermediate lobe of the pituitary	Pituitary stalk	Hypothalamus

Connects the pituitary gland to the hypothalamus	
Composed of glandular tissue	
Composed of nervous tissue	
Part of the pituitary whose function is unknown in humans	
Situated superiorly to the pituitary gland	
A supporting cell of the posterior pituitary	

Table 9.1 Anatomy of the pituitary gland

 Completion

6. Complete Table 9.2 by inserting the long-hand names of the abbreviated hormones from the hypothalamus, and the target tissues of the anterior pituitary hormone(s) that each releases.

Hypothalamus	Anterior pituitary	Target gland or tissue
GHRH –	GH	
GHRIH –	GH Inhibition TSH Inhibition	
TRH –	TSH	
CRH –	ACTH	
PRH –	PRL	
PIH –	PIH	
LHRH –	FSH	
also known as GnRH –	LH	

Table 9.2 Hormones of the hypothalamus, anterior pituitary and their target tissues

7. Complete Table 9.3 by adding the full names and functions of anterior pituitary hormones.

Hormone	Abbreviation	Function
	GH	
	TSH	
	ACTH	
	PRL	
	FSH	Males: Females:
	LH	Males: Females:

Table 9.3   Summary of the hormones secreted by the anterior pituitary gland

 ## Labelling and completion

8. Label the structures identified in Figure 9.2.

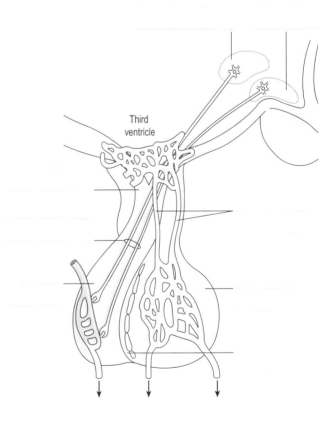

Third ventricle

Figure 9.2   The pituitary gland

9. Name the two hormones secreted by the posterior pituitary: _____   _____.

10. The anterior lobe of the pituitary gland is also known as the: _____.

11. The posterior lobe of the pituitary gland is also known as the: _____.

Completion

12. Complete the boxes labelled a and b in Figure 9.3.

13. Match the hormones and effects from the key choices with the numbered parts of Figure 9.3.

Key choices:
Trophic hormones
Releasing hormones
Lowered
Raised

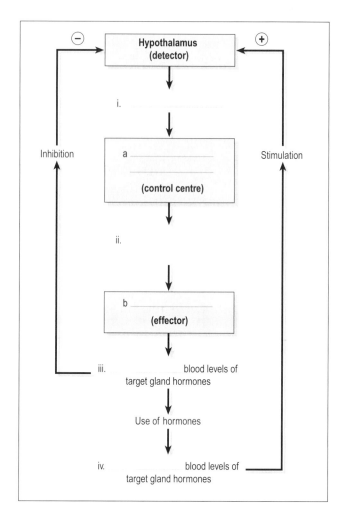

Figure 9.3 Negative feedback regulation of blood hormone levels

 MCQs

14. Secretion of which hormones normally increases during sleep (choose all that apply): _____.

    a. Growth hormone
    b. Thyroid stimulating hormone

    c. Prolactin
    d. Luteinizing hormone.

15. The hormone that promotes lactation is: _____.

    a. Melatonin
    b. Thymosin
    c. Testosterone
    d. Prolactin.

16. During childhood there is very limited secretion of: _____.

    a. Luteinizing hormone releasing hormone
    b. Growth hormone

    c. Thyroid stimulating hormone
    d. Melatonin.

17. Circadian rhythm means that regular fluctuations in hormone levels occur over a period of: _____.

    a. An hour
    b. A day
    c. A week
    d. A year.

18. T_4 is also known as: _____.

    a. Thyroid stimulating hormone
    b. Thyroglobulin

    c. Thyroxine
    d. Tri-iodothyronine.

Completion

19. Complete the paragraph below to describe the secretion and effects of antidiuretic hormone (ADH).

An increase in the rate of urine production is called _____. ADH is secreted by the _____

pituitary gland; its main effect is to _____ urine output. It does this by _____ the permeability

of the _____ convoluted tubules and the _____ ducts in the nephrons of the kidneys to

_____. As a result, more water is reabsorbed from the filtrate. Secretion of ADH occurs in response to

increasing _____ of the blood, which is detected by _____-receptors in the hypothalamus.

Situations where this takes place include _____ and _____ – more water is reabsorbed

decreasing the blood _____. In more serious situations, ADH also causes _____ of smooth

muscle, which results in _____ in small arteries. This has a pressor effect, increasing _____

pressure, which reflects the alternative name of this hormone, _____.

 Matching

20. Complete the paragraph below using the key choices listed, to provide an account of the effects of oxytocin:

Key choices:
Myoepithelial cells
Stimulation
Hypothalamus
Lactation
Parturition
Positive
Posterior pituitary
Stimulates
Stretch receptors
Uterine cervix
Smooth muscle

Oxytocin stimulates two target tissues before and after childbirth. These are

uterine _____ and _____ of the lactating

breast. During childbirth, also known as _____, increasing amounts of

oxytocin are released in response to increasing _____ of sensory

_____ in the _____ by the baby's head. Sensory

impulses are generated and travel to the _____, stimulating the

_____ to secrete more oxytocin. This _____ the

uterus to contract more forcefully, moving the baby's head further downwards

through the uterine cervix and vagina. The mechanism stops shortly after the

baby has been born. This is an example of a _____ feedback mechanism.

After birth oxytocin stimulates _____.

THYROID GLAND

 Matching

21. Match the key choices from the list with the statements in Table 9.4 to describe the structure of the thyroid gland.

Key choices: Iodine Isthmus Parathyroid glands Parafollicular cells Thyroglobulin TSH TRH T$_4$ Recurrent laryngeal Capsule	

The thyroid gland is surrounded by this structure	
Joins the two thyroid lobes together	
Lie against the posterior surface of the thyroid gland	
Secrete the hormone calcitonin	
Constituent of T$_3$ and T$_4$	
Secreted by the hypothalamus	
Thyroxine	
Precursor of T$_3$ and T$_4$	
Secreted by the anterior pituitary	
The nerves close to the thyroid gland	

Table 9.4   Features of the thyroid gland

  Colouring and matching

22. Colour and match the parts of the thyroid gland shown in Figure 9.4:

- ○ Blood vessels
- ○ Follicles
- ○ Follicular cells
- ○ Interlobular connective tissue
- ○ Parafollicular cells

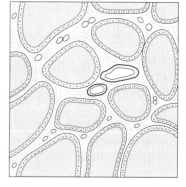

Figure 9.4   The microscopic structure of the thyroid gland

Pot luck

23. Identify whether the statements below are TRUE (T) or FALSE (F).

    a. The main source of iodine is seafood. _____

    b. Calcitonin is secreted by the parathyroid glands. _____

    c. Thyroxine is secreted by the anterior pituitary. _____

    d. Thyroid hormones are stored in the thyroid follicles until stimulation by thyroid releasing hormone occurs. _____.

Completion

24. Complete Table 9.5 to summarize the effects of excess and deficiency of T_3 and T_4.

Body function affected	Hypersecretion of T_3 and T_4	Hyposecretion of T_3 and T_4
Metabolic rate		
Weight		
Appetite		
Mental state		
Scalp		
Heart		
Skin		
Faeces		
Eyes		None

Table 9.5 Effects of abnormal secretion of thyroid hormones

PARATHYROID GLANDS

  Colouring and matching

25. Colour and match the structures indicated on Figure 9.5:

- ○ Arteries
- ○ Veins
- ○ Oesophagus
- ○ Parathyroid glands
- ○ Thyroid gland
- ○ Pharynx
- ○ Recurrent pharyngeal nerves

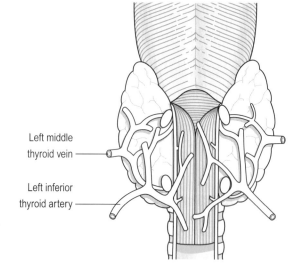

Left middle thyroid vein —

Left inferior thyroid artery —

Figure 9.5 The positions of the parathyroid glands and their related structures, viewed from behind

26. Name the hormone secreted by the parathyroid glands. _____.

27. Name the hormone secreted by the thyroid gland which acts in a complementary way to the one above.

_____.

 Pot luck

28. Identify the *six* mistakes in the paragraph below and correct them.

The parathyroid glands secrete parathyroid hormone (PTH) and blood calcium levels regulate its secretion. When they rise, secretion of PTH is increased and vice versa. The main function of PTH is to decrease the blood calcium level. This is achieved by decreasing the amount of calcium absorbed from the small intestine and reabsorbed from the renal tubules. If these sources do not provide sufficient levels then PTH stimulates osteoblasts (bone-destroying cells) and calcium is released into the blood from the parathyroid glands. Normal blood calcium levels are needed for muscle relaxation, blood clotting and nerve impulse transmission.

ADRENAL GLANDS

 Matching

29. Match the key choices below with the statements in Table 9.6 to complete characteristics of the adrenal glands.

Key choices:	
Aldosterone	Kidneys
Androgens	Suprarenal
Medulla	Hydrocortisone
Cortex	Cholesterol

Is essential for life	
Inner part of the adrenal gland	
Veins that drain the adrenal glands	
The organs immediately inferior to the adrenal glands	
Male sex hormones	
The lipid that forms the basic structure of adrenocorticoids	
A mineralocorticoid hormone	
A glucocorticoid hormone	

Table 9.6   Features of the adrenal glands

Completion

30. Identify the structures labelled a, b, c and d on Figure 9.6.

31. Insert an arrow in each circle on Figure 9.6 to indicate whether the response to stress is to increase or decrease each effect.

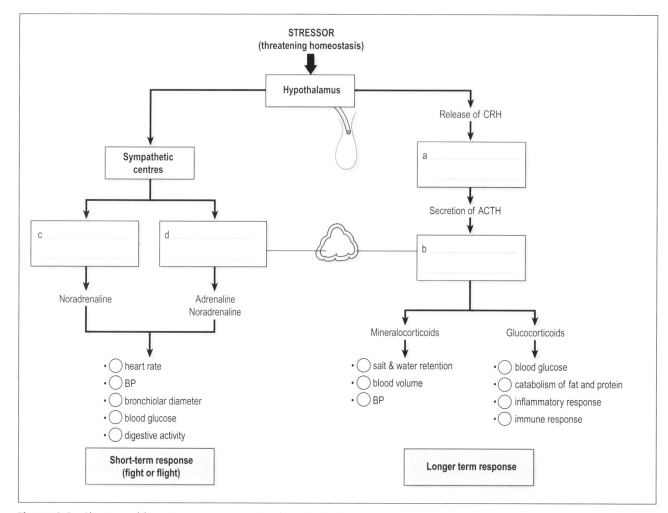

Figure 9.6 Short- and long-term responses to physiological stressors

PANCREATIC ISLETS

 Pot luck

32. Name the pancreatic cells that secrete these hormones:

   a. Insulin: _____

   b. Glucagon: _____

   c. Somatostatin: _____.

33. What other name is used for somatostatin? _____.

34. State whether each statement below is TRUE (T) or FALSE (F):

a. Insulin is formed from amino acids: _____

b. Normal blood glucose levels range from 6.1 to

9.9 mmol/litre: _____

c. Insulin reduces blood glucose levels: _____

d. Glucagon reduces blood sugar levels: _____

e. Secretion of insulin is stimulated by low blood

sugar levels: _____

f. Secretion of insulin is stimulated by gastrin:

_____

g. The hypothalamus is involved in secretion of

insulin: _____

h. Insulin secretion is decreased by sympathetic

stimulation: _____ .

Completion

35. Enter the effect of each pathway on metabolism in the middle column of Table 9.7.

36. In the right-hand column enter the hormone that stimulates the pathway – insulin or glucagon.

Metabolic pathway	Effect of pathway on metabolism	Stimulated by insulin or glucagon?
Gluconeogenesis		
Lipogenesis		
Glycogenesis		
Glycogenolysis		
Lipolysis		

Table 9.7   The effect of insulin and glucagon on metabolic processes

LOCAL HORMONES

? MCQs

37. The pineal gland is located: _____
    a. On the floor of the third ventricle in the brain
    b. In the fourth ventricle of the brain
    c. Below the hypothalamus in the brain
    d. In the upper part of the mediastinum behind the sternum.

38. Secretion of which hormone below is associated with a circadian rhythm and is highest at night? _____ .

   a. Prostaglandins   b. Histamine   c. Melatonin   d. Thymosin.

39. Which local hormones are involved in platelet aggregation (choose all that apply)? _____ .

   a. Histamine   b. Prostaglandins   c. Serotonin   d. Secretin.

40. Histamine has a role in: (choose all that apply) _____.

    a. The inflammatory process
    b. Control of body temperature

    c. Regulating blood pressure
    d. Secretion of gastric juice.

41. Which of the following is true of prostaglandins: _____.

    a. They inhibit labour, maintaining pregnancy
    b. They are involved in resetting the body thermostat during fever

    c. They are lipid-based
    d. They are long acting.

The respiratory system

The respiratory system is a collection of tissues and organs whose collective function is primarily oxygen intake and carbon dioxide elimination. Conventionally, the respiratory system is divided into the upper respiratory tract (those structures not contained within the chest) and the lower respiratory tract (those structures found inside the chest).

  Labelling and matching

1. Label the following parts of the respiratory system on Figure 10.1, using the following terms:

Nasal cavity	Base of left lung	Right secondary bronchus
Pharynx	Parietal pleura	Right primary bronchus
Epiglottis	Visceral pleura	Ribs
Larynx	Trachea	Thyroid cartilage
Apex of right lung	Left primary bronchus	Heart space

  Colouring and matching

2. Colour and match the:

- ○ Diaphragm
- ○ Pleural cavity
- ○ Clavicles
- ○ Tracheal cartilages

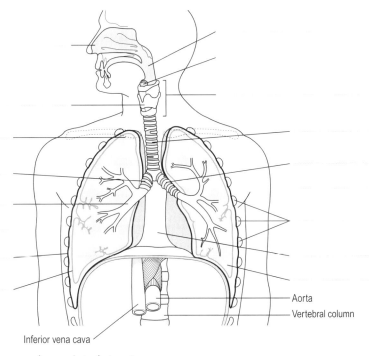

Figure 10.1 The respiratory system and associated structures

THE UPPER RESPIRATORY TRACT

 ### Matching

3. Match the structures below with their functions.

Functions	*Structures*
a. Increases surface area to moisten and warm air	Nasopharyngeal tonsil
b. Contains the vocal cords	Epiglottis
c. Forms the Adam's apple	Nasal conchae
d. The lid of the larynx, protecting the tracheal opening	Thyroid cartilage
e. Muscle flap in the roof of the mouth	Larynx
f. A collection of lymphoid tissue, involved in immunity	Auditory tube
g. Links the nasopharynx and middle ear	Soft palate

 ### Colouring and labelling

4. Colour and label the structures in Figure 10.2.

- ○ Cricoid cartilage
- ○ Epiglottis
- ○ Thyroid cartilage
- ○ Hyoid bone
- ○ Thyrohyoid membrane
- ○ Rings of tracheal cartilage

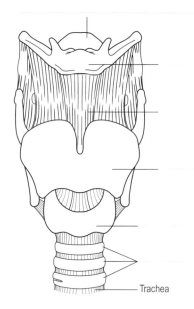

Trachea

Figure 10.2 Larynx viewed from the front

 ### Completion

5. Complete the following paragraph by inserting the correct word(s) in the spaces provided.

The upper respiratory passages carry air in and out of the respiratory system, but they have other functions too. The cells of their mucous membrane have _____, tiny hair-like structures that _____ in a wave-like motion towards the _____. They carry _____, which has been made by the _____ cells in the epithelial layer, and which traps _____ and _____ on its sticky surface. The air is therefore _____ by these mechanisms before it gets into the lungs. As the air passes through the nasal cavity, it is also _____ and _____ as it passes over the nasal _____, bony projections covered in mucous membrane. The nasal cavity also contains _____, which is covered in _____, and acts as a coarse filter for the air passing through. Immune tissue is present in patches called _____, which make _____ and therefore protect against inhaled antigens. Not only air passes through the pharynx, but also _____ and _____, and the tracheal opening is barricaded against these by the _____.

Labelling

6. Label the structures shown on Figure 10.3.

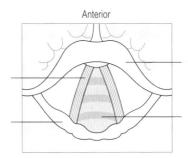

Anterior

Figure 10.3 Interior of the larynx viewed from above

 MCQs

7. Which of the following laryngeal cartilages is paired? _____

   a. Epiglottis   **b.** Thyroid cartilage   **c.** Cricoid cartilage   **d.** Arytenoid cartilage.

8. Which of the following statements about the larynx is true? _____

   a. It opens into the laryngopharynx above and the trachea below
   b. During speech, the epiglottis closes over the laryngeal opening
   c. It lies immediately anterior to the thyroid gland
   d. It is lined with non-ciliated epithelium, to allow speech.

9. Olfactory epithelium: _____

   a. Is found in the upper part of the roof of the nasal cavity
   b. Contains mainly lymphoid tissue
   c. Produces mucus to trap inhaled particles
   d. Lines the vocal cords for protection.

10. Which of the following is NOT a function of the pharynx? _____

    a. Warming of inspired air
    b. Protection from infection, by the tonsils in its walls
    c. Closing the glottis during swallowing
    d. Providing a passageway for food going to the stomach.

 Pot luck

11. Four of the five statements below, which describe speech, are incorrect. Identify the incorrect statements and write the correct versions in the spaces provided:

    a. The vocal cords in the larynx are made of bands of smooth muscle stretched across the laryngeal lumen.

    _____

    b. Increasing the speed of vibration of the vocal cords increases the pitch of sound produced.

    _____

c. Males usually have lower pitched voices than females because their vocal cords are generally longer. _____

_____

d. At rest, the vocal cords are adducted (open) to keep the laryngeal lumen clear. _____

_____ .

e. Sound is produced by air passing through the larynx on its way to the lungs, vibrating the vocal cords as it

does so. _____ .

THE LOWER RESPIRATORY TRACT

Colouring and matching

12. On Figure 10.4, colour and match the following structures:

○ Oesophagus
○ Trachea
○ Tracheal cartilage rings
○ Trachealis muscle

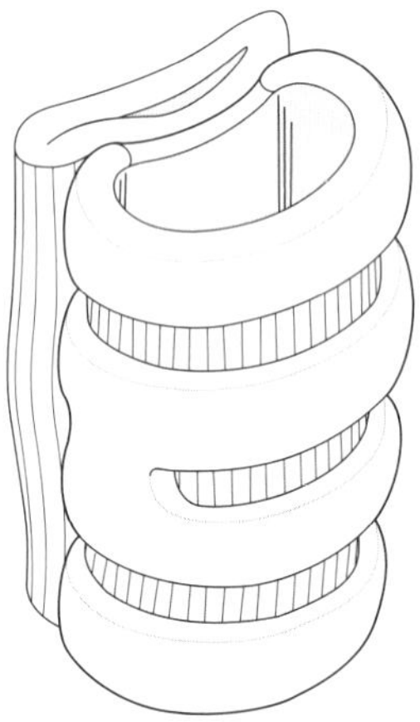

13. What shape are the tracheal cartilage rings?

_____ .

14. Why are they this shape?

_____

_____ .

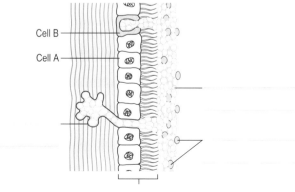

Figure 10.4 The relationship of the trachea to the oesophagus

Labelling

15. Label the structures indicated on Figure 10.5.

16. There are two types of cell in this epithelial layer; identify them and state their functions:

Cell A: _____

_____ .

Cell B: _____

_____ .

Cell B ─────
Cell A ─────

Figure 10.5 Microscopic view of ciliated mucous membrane

Matching

17. For each of the four statements in list A, identify its most appropriate reason from list B. (You won't need all the items in list B.)

List A

Mucus is produced in the upper respiratory tract because...

_____

_____

Cilia are present in the upper respiratory tract because...

_____

_____

Cartilage is present in the upper respiratory tract because...

_____

_____

Elastic tissue is present in the upper respiratory tract because...

_____

_____

List B

... the passageway has to be flexible to allow head and neck movement

... the oesophagus is normally collapsed

... the oesophagus needs to expand during swallowing

... mucus needs to be swept away from the lungs

... the tissues need to return to their original shape

... this is an efficient way of removing dust and dirt from inhaled air

... mucus builds up during normal respiration

... the airways have to be kept open at all times

... inspired air must be warmed and humidified.

 Pot luck

18. In the following paragraph, which describes the respiratory tree, there are *five* inaccuracies. Find them and correct them.

Ciliated respiratory epithelium lines the entire respiratory tree from the trachea to the alveoli, and its job is to keep the lungs clean. Cartilage rings support the airway walls; as the airways progressively divide and their diameter decreases, the amount of cartilage present increases. The smallest airways are called respiratory bronchioles, although no gas exchange takes place across their walls. The airways terminate in clusters of microscopic pouches called alveoli; it is here that most gas exchange takes place. The alveolar walls are only one cell thick and contain macrophages, which make surfactant to keep the alveoli from collapsing. Gas exchange occurring across the alveolar walls is called internal respiration.

Colouring, matching and labelling

19. In Figure 10.6, colour and label the structures indicated.

20. Colour and match the lobes of the lungs.

Right lung	Left lung
Superior lobe	Superior lobe
Middle lobe	Inferior lobe
Inferior lobe	

21. Name the area between the lungs, in which the heart and other structures lie.

_____ .

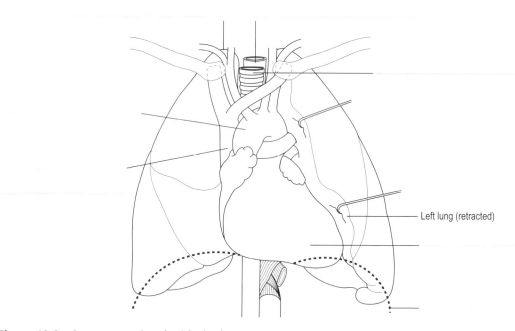

Left lung (retracted)

Figure 10.6 Organs associated with the lungs

Labelling

22. Label the visceral pleura, the parietal pleura, the pleural cavity, the diaphragm and the hilum of the right lung on Figure 10.7.

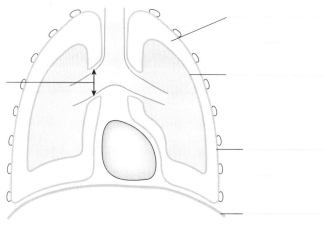

Figure 10.7 Relationship of pleura to the lung

Pot luck

23. From list A, identify all the items describing the function of pleural fluid.

_____

_____ .

List A

a. Prevents friction between the beating heart and the lungs

b. Is produced by the pleural membranes

c. Assists in effective gas exchange at the respiratory membrane

d. Reduces surface tension in the alveoli

e. Is present in a volume of about half a litre

f. Reduces friction of the lungs expanding in the thoracic cavity

g. Has antibacterial properties

h. Is produced by cilia.

 MCQs

24. What is the carina? _____

    a. One of the cartilages of the larynx
    b. The area of the lung where the airways enter
    c. The point at which the trachea divides in two
    d. The hollow of the left lung accommodating the heart.

25. Coughing: _____

    a. Occurs during inspiration
    b. Is controlled by the voluntary nervous system
    c. Is preceded by a deep inspiration
    d. Requires relaxed abdominal muscles.

26. The diameter of bronchioles is: _____

    a. Controlled by bands of skeletal muscle in their walls
    b. Increased by the parasympathetic nervous system
    c. Mainly regulated by the endocrine system
    d. Decreased by their cartilage plates.

27. Which of the following describes the function of surfactant? _____ .

    a. Increases surface tension in the alveoli
    b. Lubricates the lungs during breathing
    c. Protects the alveoli from infection
    d. Prevents the alveoli from collapsing.

Labelling, matching and colouring

28. As the respiratory tree progressively divides, the passageways become narrower and narrower. Label the structures on Figure 10.8. Indicate by colouring the large arrows the sections of the respiratory tree important in:

○ Air conduction
○ Gas exchange

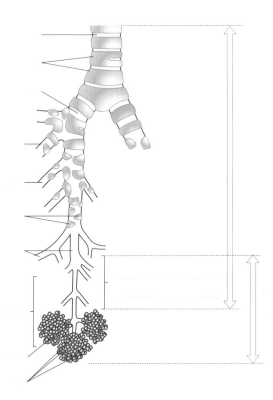

Figure 10.8 Lower respiratory tract

29. How many alveoli are there in each lung? _____.

30. As the tiniest bronchioles and alveoli have no cartilage, what prevents them from collapse?

_____.

31. On Figure 10.9, colour and match the:

○ Alveolar endothelial cells
○ Elastic connective tissue
○ Blood capillaries

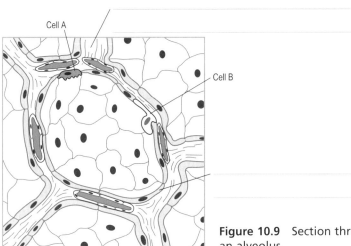

Cell A

Cell B

Figure 10.9 Section through an alveolus

32. Regarding Figure 10.9, complete the following sentences regarding cells A and B.

Cell A produces the substance that provides

an oily lining for the alveolus; this substance

is called _____ and the cell is a _____ cell. Cell B is involved

in protection; it cleans the alveolus by the process of _____ ; it is a _____.

Pot luck

33. Put the following statements in order, so that the flow of blood through the heart, lungs and systemic circulation is correctly summarized. The first one has been done to start you off.

Right ventricle	Body tissues
Pulmonary vein	Aorta
Pulmonary artery	Left ventricle
Lungs	Right atrium

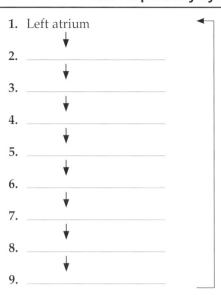

1. Left atrium
2. _____
3. _____
4. _____
5. _____
6. _____
7. _____
8. _____
9. _____

RESPIRATION

Completion

34. The paragraphs below describe a normal cycle of respiration. Fill in the blanks, using the terms supplied:

Passive	Increases	Inflate	Upwards
Deflate	Outwards	Muscular effort	Downwards
Inwards	Downwards	Into	External intercostal muscles
Relaxed	Decreases	Contracts	Increases
Relaxes	External intercostal muscles	Out of	Decreases

Just before inspiration commences, the diaphragm is _____; this occurs in the pause between breaths in normal quiet breathing. Inspiration commences. The ribcage moves _____ and _____ owing to contraction of the _____. The diaphragm _____ and moves _____. This _____ the volume of the thoracic cavity, and _____ the pressure. Because of these changes, air moves _____ the lungs, and the lungs _____. Inspiration has taken place.

Unlike inspiration, expiration is usually a _____ process because it requires no _____. So, following the end of inspiration, the diaphragm _____ and moves back into its resting position. The ribcage moves _____ and _____, because the _____ have relaxed. This _____ the volume of the thoracic cavity, and so _____ the pressure within it. Air therefore now moves _____ the lungs, and they _____. There is now a rest period before the next cycle begins.

Matching

35. Match each of the following descriptions of lung tissue with the following terms: elastic (E), compliant (C), both (C/E) or neither (N).

   a. Easily stretched but does not return to its original shape _____

   b. Easily stretched and returns to its original shape _____

   c. Resistant to stretch and does not return to its original shape _____

   d. Resistant to stretch and returns to its original shape _____.

36. Which of the descriptions best describes the healthy lung? _____.

Labelling

37. Complete the list below, identifying each of the standard abbreviations for the main lung volumes and capacities, and use the abbreviations to label Figure 10.10:

   a. TV _____

   b. VC _____

   c. IC _____

   d. RV _____

   e. IRV _____

   f. ERV _____

   g. TLC _____.

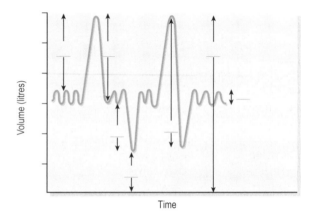

Figure 10.10 Lung volumes and capacities

Applying what you know

38. If the functional residual capacity is 3000 ml, the tidal volume is 500 ml and the total lung capacity is 6000 ml, calculate the inspiratory capacity and the inspiratory reserve volume.

_____.

39. If the total lung capacity is 6000 ml and the residual volume is 1200 ml, what is the vital capacity?

_____.

40. Calculate the alveolar ventilation for an individual whose tidal volume is 450 ml, anatomical dead space is 160 ml and respiratory rate is 13/min.

_____.

41. You are asked to calculate a person's total lung capacity. You are given the values for tidal volume, inspiratory reserve volume and residual volume. Which additional value will you need to complete the calculation?

_____ .

42. Emily is on the treadmill in the gym. Her pulse is 140/min, the tidal volume is 1200 ml, and her respiratory rate is 20/min. Of the volumes and capacities labelled in Figure 10.9, which two would be unchanged if you were to measure them right now? Explain your answers.

_____

_____

_____

_____ .

Definitions

Define the following terms:

43. External respiration:

_____

_____

_____ .

44. Internal respiration:

_____

_____

_____ .

Pot luck

45. List two features of the alveolar membrane that increase efficiency of gas exchange:

- _____

- _____ .

46. List two features of the blood flow through the alveolar capillaries that increase efficiency of gas exchange:

- _____

- _____ .

MCQs

47. The composition of atmospheric air is: _____

    a. 38% oxygen, 60% nitrogen and 5% other gases
    b. 21% oxygen, 78% hydrogen and 1% other gases
    c. 38% nitrogen, 60% oxygen and 5% other gases
    d. 21% oxygen, 78% nitrogen, and 1% other gases.

48. Air in the alveoli: _____

    a. Is similar in composition to atmospheric air
    b. Contains twice as much oxygen as atmospheric air
    c. Is at a lower pressure than atmospheric air
    d. Is fully saturated with water vapour.

49. External respiration: _____

    a. Stops during expiration
    b. Is continuous throughout the respiratory cycle
    c. Occurs mainly during inspiration
    d. Stops during breath-holding.

50. Oxygen diffusion across the respiratory membrane is increased by: _____.

    a. Oxygen pumps in the alveolar wall
    b. An intact surfactant layer
    c. A concentration gradient for oxygen between the alveoli and the blood
    d. Low levels of carbon dioxide in the alveoli.

  Colouring and completion

51. Figure 10.11 shows gas exchange between an alveolus and a lung capillary.

    What is this process called?

    _____.

52. Using different colours for carbon dioxide and oxygen, colour in the arrows to show how each gas moves.

53. Complete the boxes to show the partial pressures of each gas in the arterial capillary, the venous capillary and the alveolus.

54. Colour the region on Figure 10.11 that represents the respiratory membrane.

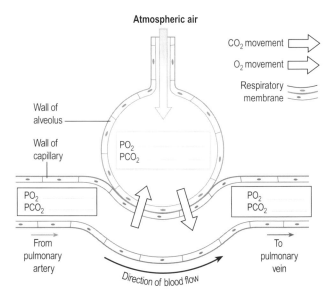

Figure 10.11 Gas exchange between alveoli and bloodstream

55. Figure 10.12 shows gas exchange between the bloodstream and tissue cells.

What is this process called? _____.

56. Using different colours for carbon dioxide and oxygen, colour in the arrows to show how each gas moves.

57. Complete the boxes to show the partial pressures of each gas in the arterial capillary, the venous capillary and the tissue cells.

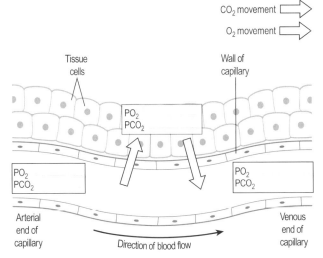

Figure 10.12 Gas exchange between the bloodstream and tissue cells

 Pot luck

58. In each sentence in the box below, the initial statement is true. In *four* of the sentences, the reason given does not explain the initial statement. Identify the four inaccurate reasons and correct the explanation.

Statement	Name of nerve	Reason
A. Carbon dioxide diffuses from the body tissues into the bloodstream	because	PCO_2 is lower in the tissues than in the bloodstream.
B. Tissue levels of oxygen are lower than capillary blood levels	because	body cells are constantly producing carbon dioxide.
C. Oxygen diffuses out of the capillary into the tissue cells	because	the tissues require a constant supply of oxygen.
D. Oxygen levels at the venous end of the capillary are lower than the arterial end	because	oxygen diffuses into the tissues as the blood flows through the capillary.
E. Carbon dioxide moves out of the pulmonary capillaries into the alveoli	because	it is being exchanged for oxygen across the respiratory membrane.

A. _____

B. _____

C. _____

D. _____

E. _____.

TRANSPORT OF GASES

 Pot luck

59. Decide whether the following statements apply to carbon dioxide, to oxygen, or to both.

   a. Waste product of metabolism: _____

   b. 23% carried bound to haemoglobin: _____

   c. Raised temperatures increase release from haemoglobin: _____

   d. Mainly carried as bicarbonate ions in the plasma: _____

   e. 98.5% carried bound to haemoglobin: _____

   f. Binds loosely to haemoglobin: _____

   g. Binding with haemoglobin is tighter in the lungs than in the tissues: _____

   h. 1.5% carried dissolved in plasma: _____

   i. Binding to haemoglobin is tighter in the tissues than in the lungs: _____.

CONTROL OF RESPIRATION

  Labelling and colouring

60. Label the structures indicated on Figure 10.13 using the items listed. Colour the nerve supply to the muscles of respiration.

| Respiratory rhythmicity centre in medulla oblongata |
| Cerebral cortex |
| Glossopharyngeal nerve |
| Spinal cord |
| Intercostal muscles |
| Diaphragm |
| Intercostal nerves |
| Phrenic nerve |
| Carotid body |

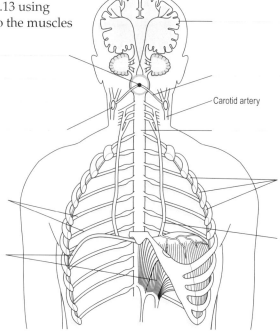

Carotid artery

Figure 10.13 Some of the nerves involved in control of respiration

Matching

61. State whether the following would increase (I) or decrease (D) the rate and depth of respiration.

a. Decreased blood hydrogen ion concentration _____   f. Stimulation of the Hering–Breuer reflex _____

b. Rising blood carbon dioxide levels _____   g. Sleep _____

c. Falling blood pressure _____   h. Rising blood oxygen levels _____

d. Anxiety _____   i. Decreased pH of the cerebrospinal fluid _____

e. Increased sympathetic activity _____   j. Hypothermia _____.

? **MCQs**

62. Which one of the following is true of the chemical control of respiration? _____

a. Chemoreceptors are found only in the brain
b. Chemoreceptors principally detect falling O_2 levels
c. Central chemoreceptors are found in the cerebral cortex
d. A fall in cerebrospinal fluid pH stimulates central chemoreceptors.

63. Which of the following are true (choose all that apply)? _____

a. The respiratory centre is located in the brainstem
b. Accessory muscles of respiration include the diaphragm and the sternocleidomastoid
c. Stimulation of the vagus nerve supplying aortic chemoreceptors increases the activity of the respiratory centre
d. Control of breathing is entirely involuntary.

64. The Hering–Breuer reflex controls respiration by measuring: _____

a. Arterial blood pressure   c. CO_2 levels in the cerebrospinal fluid
b. Airway resistance   d. Stretch in the lungs.

65. Which of the following is the key to why a child in a tantrum cannot hold his or her breath indefinitely? _____.

a. Rising blood CO_2 levels   c. Rising blood pH
b. Falling blood O_2 levels   d. Falling blood $[H^+]$.

Definitions

66. Define the following terms:

a. Hypercapnia _____

b. Hypoxia _____

c. Hypoxaemia _____

d. Carbaminohaemoglobin _____

e. Oxyhaemoglobin _____.

Nutrition

All body cells need a supply of nutrients in appropriate quantities and the ultimate source of these nutrients is the diet. This chapter considers the main groups of nutrients and their roles in body function.

 Pot luck

1. List the main nutrient groups needed for a balanced diet:

   - _____

   - _____

   - _____

   - _____

   - _____.

2. Calculate the body mass index (BMI) for the following individuals:

   a. A man who is 1.9 m tall and weighs 60 kg _____

   b. A man who is 1.8 m tall and weighs 90 kg _____

   c. A woman who 1.6 m tall and weighs 50 kg _____

   d. A woman who is 1.7 m tall and weighs 90 kg _____.

3. Identify whether the BMI of each person above is underweight, normal, overweight or obese.

  Labelling and matching

4. Each section in Figure 11.1 represents a food group and its recommended proportions for a healthy diet. Label each of them, using the key choices below.

> _Key choices:_
> Milk and dairy products
> Bread, rice, potatoes and pasta
> Fruit and vegetables
> Foods and drinks high in fat and/or sugar
> Meat, fish, eggs and beans

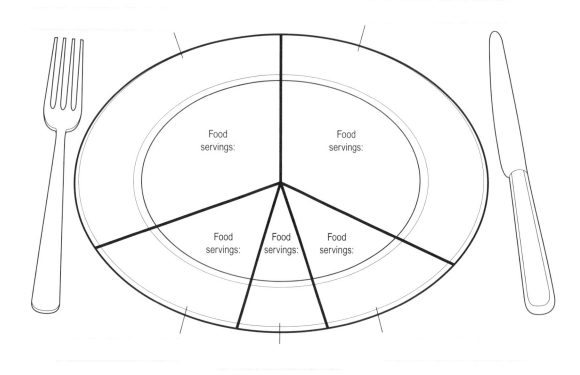

Figure 11.1 The main food groups and recommended proportions within a balanced diet

CARBOHYDRATES, PROTEINS AND FATS

? MCQs

5. The correct units for expressing dietary energy are: _____.

   a. Grams (g)　　　b. Millilitres (ml)　　　c. Kilograms (kg)　　　d. Kilojoules (kJ).

6. Which of the following correctly describes the chemical composition of carbohydrates? _____.

   a. Carbon, hydrogen, oxygen, nitrogen, sulphur with hydrogen and oxygen being in the same proportion as water
   b. Carbon, hydrogen, oxygen, nitrogen, with hydrogen and oxygen being in any proportion
   c. Carbon, hydrogen, oxygen, with hydrogen and oxygen being in any proportion
   d. Carbon, hydrogen, oxygen, with hydrogen and oxygen being in the same proportion as water.

7. Which of the following are digestible sugars (choose all that apply)? _____.

   a. Cellulose　　　b. Monosaccharides　　　c. Disaccharides　　　d. Polysaccharides.

8. Excess dietary carbohydrate can be stored as (choose all that apply): _____.

   a. Glycogen　　　b. Glucose　　　c. Fat　　　d. Protein.

Definitions

Define the following terms:

9. A balanced diet _____

_____.

10. Essential amino acid _____

_____.

11. Non-essential amino acid _____ _____

_____.

12. Biological value of protein _____

_____.

Pot luck

13. List the elements always present in amino acids _____.

14. Name some minerals which can also be constituents of amino acids _____

_____.

15. Outline the main functions of amino acids in the body:

   • _____

   • _____

   • _____.

Completion

16. The following paragraph discusses the structure and function of the fats. Complete it by filling in the blanks.

The three elements that make up fat are _____, _____ and _____. Fats are

usually divided into two groups: _____ fats are found in foods from animal sources, such as

_____, _____ and _____. The second group, the _____ fats,

are found in vegetable oils. Fat (adipose) tissue is laid down under the skin, where it acts as a(n)

_____. It is also found around the kidneys, where its function is to _____ these organs.

Fat depots in the body are important as _____ sources. Certain hormones, such as

_____ e.g. cortisone, are synthesized from the fatty precursor _____, also found in the

cell membrane. In addition, certain substances are absorbed with fat in the intestine, a significant example being the _____, which are essential for health despite being required only in very small amounts. Fats in a meal have the direct effect of _____ gastric emptying and _____ the return of a feeling of hunger.

THE VITAMINS

17. Table 11.1 lists the main vitamins. Complete it by filling in the main dietary sources of each vitamin.

Vitamin	Main sources
A	
B$_1$ (thiamine)	
B$_2$ (riboflavine)	
Folate (folic acid)	
Niacin	
B$_6$ (pyridoxine)	
B$_{12}$ (cyanocobalamin)	
Pantothenic acid	
Biotin	
C	
D	
E	
K	

Table 11.1   Vitamin sources

18. Underline the fat soluble vitamins in Table 11.1.

Matching

19. Vitamins act as cofactors in a range of important biochemical reactions in the body. Assign to each of the functions in list A the appropriate vitamin from list B. (You may need the items in list B more than once, and you can use more than one vitamin for each function.)

List A

a. Antioxidant: _____

b. Connective tissue synthesis: _____

c. Manufacture of visual pigments: _____

d. Non-essential amino acid synthesis: _____

e. Cell growth and differentiation, especially fast growing tissues: _____

f. Carbohydrate metabolism: _____

g. Synthesis of clotting factors: _____

h. DNA synthesis: _____

i. Amino acid/protein metabolism: _____

j. Regulation of calcium and phosphate levels: _____

k. Fat metabolism: _____

l. Myelin production: _____.

List B

Vitamin A

Vitamin B_1

Vitamin B_2

Vitamin B_6

Vitamin B_{12}

Folate (folic acid)

Pantothenic acid

Biotin

Niacin

Vitamin D

Vitamin E

Vitamin K

Vitamin C

20. The passage below describes the disorders that are associated with deficiency of certain vitamins. Complete it by deleting the incorrect options in bold, leaving the correct version.

Because vitamin A is a fat soluble vitamin, its absorption can be reduced if **bile/trypsin/pepsin** secretion into the gastrointestinal tract is lower than normal. The first sign of deficiency is **poor bone development/reduced immunity/night blindness**, and this may be followed by **poor blood clotting/conjunctival ulceration/ neurological symptoms**. On the other hand, the B-complex vitamins are water soluble. Most of them are involved in **repair and differentiation of tissues/maintenance of an efficient immune system/biochemical release of energy**. Thiamine deficiency is associated with **pellagra/kwashiorkor/beriberi**, and niacin inadequacy leads to **pellagra/kwashiorkor/beriberi**. Folic acid is required for **DNA/collagen/clotting factor** synthesis, and is therefore often prescribed as a supplement in pregnancy. Deficiency of vitamin B_{12} typically leads to **haemolytic/ megaloblastic/iron deficiency** anaemia, because it is needed for DNA synthesis, and is usually associated with lack of **biotin/bile/intrinsic factor** in the gastrointestinal tract.

Vitamin C is needed for **connective tissue synthesis/clotting factor synthesis/maintenance of normal bone tissue.** One of the first signs of deficiency of this vitamin is therefore loosening of the teeth, due to **defective gum tissue/bleeding into the gums because of clotting deficiency/erosion of the bony sockets.** Vitamin C is destroyed by **heat/water/low gastric pH**.

Lack of vitamin D causes **osteoporosis/osteoma/osteomalacia** in adults, and **scurvy/rickets/night blindness** in children. Vitamin E deficiency results in **haemolytic/megaloblastic/iron deficiency** anaemia, because the **cell membrane/haemoglobin content/cytoplasm** of red blood cells is damaged.

Vitamin K deficiency leads to problems with **myelination of nerves/absorption of calcium/blood coagulation.**

135

MINERALS, TRACE ELEMENTS AND WATER

 Matching

21. Complete Table 11.2 by ticking the appropriate boxes against each of the minerals shown.

	Calcium	Phosphate	Sodium	Potassium	Iron	Iodine
Needed for haemoglobin synthesis						
Used in thyroxine manufacture						
Most abundant cation outside cells						
99% of body stock is found in bones						
Most abundant cation inside cells						
May be added to table salt						
Vitamin D is needed for use						
Involved in muscle contraction						
Used to make high-energy ATP						
Needed for normal blood clotting						
Required for hardening of teeth						
Needed for normal nerve transmission						

Table 11.2 Functions of minerals

 MCQs

22. Which foods are good sources of calcium (choose all that apply)? _____.
   a. Meat   b. Biscuits   c. Cheese   d. Milk.

23. Which foods can be eaten to ensure adequate dietary phosphate (choose all that apply)? _____.
   a. Meat   b. Vegetables   c. Cheese   d. Milk.

24. Which foods contain high sodium levels (choose all that apply)? _____.
   a. Meat   b. Table salt   c. Potato crisps   d. Processed foods.

25. Which foods are good sources of potassium (choose all that apply)? _____.
   a. Meat   b. Fruit and vegetables   c. Seafood   d. Table salt.

26. Which foods provide dietary iron (choose all that apply)? _____.
   a. Liver   b. Red meat   c. Green vegetables   d. Fruit.

27. Which foods are good sources of iodine (choose all that apply)? _____.

    a. Red meat   **b.** Seafood   **c.** Vegetables   **d.** Bread.

28. People prone to dietary iron deficiency include: _____.

    a. Infants under the age of three   **b.** Pregnant women
    c. Menstruating women   **d.** Postmenopausal women.

29. Dietary sodium deficiency is: _____.

    a. Common during growth spurts
    b. Common if the diet is deficient in milk
    c. Uncommon if the diet contains enough vegetables
    d. Uncommon as it is a constituent of a wide range of foods.

30. Which of the following is true about iodine? _____.

    a. It is needed for synthesis of the thyroid hormones thyroxine and triiodothyronine
    b. A high intake may cause goitre
    c. When metabolic rate is raised, less dietary iodine is needed
    d. Meat is a common source.

31. Which of the following is true about potassium (choose all that apply)? _____.

    a. It is required for muscle contraction
    b. It is required in higher amounts in menstruating women
    c. Dietary deficiency is common especially during pregnancy
    d. Most potassium ions are inside body cells (intracellular).

Pot luck

32. What proportion of body weight is water in:

Men? _____   Women? _____.

33. Identify whether the statements below are TRUE (T) or FALSE (F).

    a. The body contains more extracellular water than intracellular water _____

    b. Extracellular water includes plasma and interstitial fluid _____

    c. Water is a constituent of faeces _____

    d. Intracellular water is found in the bladder _____

    e. Excessive water loss leads to dehydration which can have serious consequences, especially in children _____.

FIBRE

? Pot luck

34. NSP is the abbreviation for: _____.

35. List the five main functions of NSP.

- _____

- _____

- _____

- _____

- _____

36. What are the main dietary sources of NSP? _____.

37. Name the condition characterized by passing hard faeces due to inadequate dietary NSP. _____.

The digestive system

The digestive system is a varied collection of organs and tissues, which participate in some way in the digestion and absorption of food. Food and drink taken orally is not usually in a chemically appropriate form for the tissues of the body to use, and the digestive system possesses a wide array of enzymes needed to convert what we eat and drink into a form more suitable for absorption and use.

   Matching, colouring and labelling

1. Figure 12.1 shows the digestive system.
   Match and colour the main parts using the key provided.

○ Rectum	○ Duodenum
○ Liver	○ Pancreas (behind stomach)
○ Stomach	○ Oesophagus
○ Large intestine	○ Gall bladder
○ Sigmoid colon	○ Small intestine

2. Label the other structures indicated.

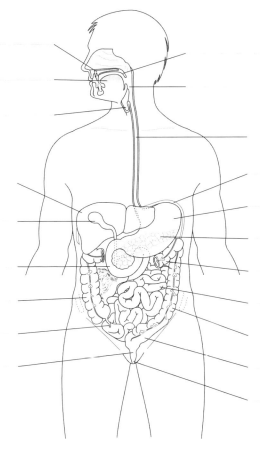

Figure 12.1 The digestive system

Pot luck

3. List the five processes that take place in the alimentary canal and briefly describe each one:

a. _____

b. _____

c. _____

d. _____

e. _____.

4. Explain the difference between mechanical and chemical digestion:

_____

_____

_____.

THE BASIC STRUCTURE OF THE GASTROINTESTINAL TRACT

Matching

5. Decide whether the following structures are classed either as parts of the alimentary tract or accessory organs of digestion, and use them to complete Table 12.1.

Mouth	Liver	Gall bladder
Parotid glands	Stomach	Oesophagus
Pancreas	Submandibular glands	Sublingual glands
Small intestine	Large intestine	Rectum and anus

Organs of alimentary tract	Accessory organs

Table 12.1 Organs of the alimentary tract and accessory organs

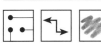 **Labelling, matching and colouring**

6. Figure 12.2 shows a section through the wall of the alimentary canal, and although the digestive organs are varied in shape and function, this basic pattern is seen in almost all regions. Colour, match and label the nerve plexuses on Figure 12.2.

○ Myenteric (Auerbach's) plexus
○ Submucosal (Meissner's) plexus

7. Label the layers shown on Figure 12.2.

8. Name the divisions of the nervous system that form networks of nerves in the myenteric plexus:

• _____

• _____ .

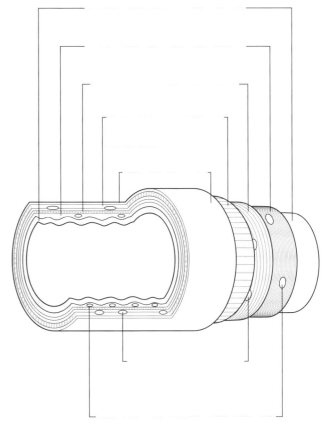

Figure 12.2 General structure of the alimentary canal

 MCQs

9. The parietal peritoneum: _____.

    a. Lines the alimentary canal
    b. Lines the abdominal wall
    c. Forms the greater omentum
    d. Covers the organs within the abdominal and pelvic cavities.

10. Which of the following is retroperitoneal? _____.

    a. Transverse colon   **b.** Liver   **c.** Stomach   **d.** Pancreas.

11. The layer of the alimentary tract involved in peristalsis is the: _____.

    a. Adventitia   **b.** Muscle layer   **c.** Submucosa   **d.** Mucosa.

12. In parts of the alimentary tract that are subject to considerable wear and tear, the mucous membrane is formed from: _____.

    a. Stratified squamous epithelium
    b. Simple squamous epithelium
    c. Columnar epithelium
    d. Transitional epithelium.

Labelling and colouring

13. What type of tissue is shown in Figure 12.3?

_____.

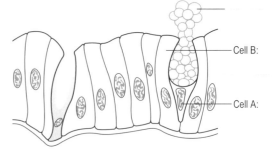

Figure 12.3   Cells of the digestive mucosa

14. Label the two cell types shown in the diagram, and colour and label the product of cell A.

15. Name two functions of the product of cell A:

- _____

- _____

16. In which regions of the digestive tract is the tissue in Figure 12.3 found? _____.

THE UPPER GASTROINTESTINAL TRACT

  Colouring and matching

17. Figure 12.4 shows structures of the mouth. Colour and match the following:

- ○ Lower lip
- ○ Upper lip
- ○ Tongue
- ○ Palatine tonsils
- ○ Palate
- ○ Uvula
- ○ Palatopharyngeal arch
- ○ Posterior wall of the pharynx

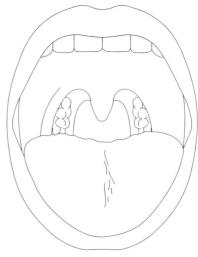

Figure 12.4   Structures of the widely open mouth

 Pot luck

18. The roof of the mouth is referred to as the palate. Compare and contrast the position and structure of the hard and soft parts.

_____

_____

_____.

Labelling and colouring

19. Figure 12.5 shows the roof of the mouth complete with teeth. Label and colour the different teeth shown.

20. State the functions of the different types of teeth in Figure 12.5.

A _____

B _____

C _____

D _____ .

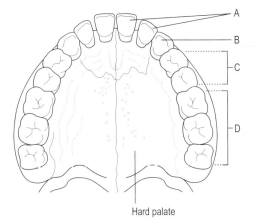

Hard palate

Figure 12.5 The roof of the mouth and the permanent teeth

 Labelling and colouring

21. Figure 12.6 shows the internal structure of a tooth. Label the structures indicated.

22. Colour the following parts on Figure 12.6 using the key, and describe the function and main characteristics of each.

○ Cement _____

_____

○ Dentine _____

_____

○ Enamel _____

_____ .

23. What is found within the pulp cavity?

_____

24. There are 32 permanent teeth and only 20 deciduous (baby) teeth. Name and number the teeth that are missing from a child's dentition.

_____

_____ .

25. Around what age does loss of the deciduous teeth begin? _____ .

26. Around what age is the permanent dentition complete? _____ .

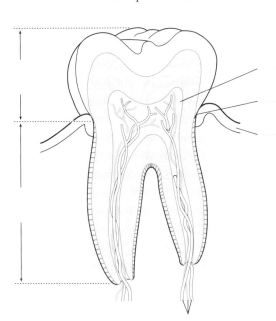

Figure 12.6 A section of a tooth

Labelling, matching and colouring

27. Figure 12.7 shows the position of the salivary glands. They are paired, so both sides of the mouth contain them. Label the structures indicated.

28. On Figure 12.7, colour and match the glands themselves using the key below and identify where the salivary glands open into the mouth.

○ _____ gland

opens into: _____

○ _____ gland

opens into: _____

○ _____ gland

opens into: _____ .

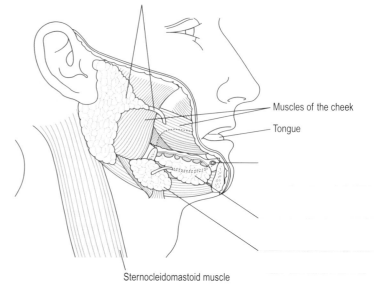

Muscles of the cheek

Tongue

Sternocleidomastoid muscle

Figure 12.7 Position of salivary glands

 Labelling

29. Figure 12.8 shows the stomach and structures close by. Label the structures indicated.

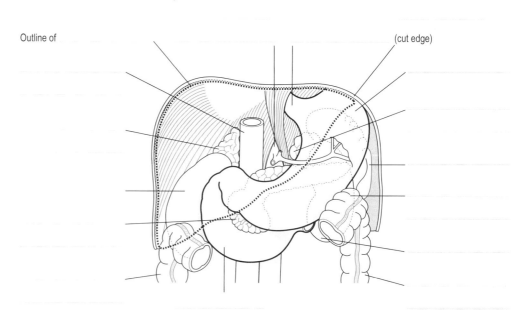

Outline of

(cut edge)

Figure 12.8 Stomach and associated structures

30. Figure 12.9 shows the main parts of the stomach. Label the structures indicated.

✎ Completion

31. The stomach has three layers of smooth muscle.

The inner layer consists of _____ fibres,

the middle layer consists of _____ fibres

and the outer layer of _____ fibres.

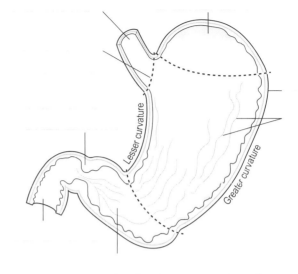

Figure 12.9 Longitudinal section of the stomach

❓ Pot luck

32. What is the functional importance of the three layers of smooth muscle? _____

_____ .

33. Describe the structure and function of a sphincter.

_____

_____ .

34. List A gives some of the important constituents of gastric juice. Match each with the relevant statements in list B. (You will need some items in list A more than once.)

List A
Hydrochloric acid
Intrinsic factor
Mucus
Pepsinogens.

List B

a. Secreted by parietal cells: _____

b. Provides correct pH for stomach enzymes: _____

c. Lubricates gastric contents: _____

d. Kills ingested bacteria: _____

e. Activated to form pepsins: _____

f. Secreted by chief cells: _____

g. Inactive enzymes: _____

h. Required for absorption of vitamin B_{12} in the ileum: _____

i. Gives a pH of 1–3 in the stomach: _____

j. Secreted by goblet cells: _____

k. Precursor for protein digestion: _____

l. Stops the action of salivary amylase: _____

? MCQs

35. In relation to the stomach, the left lobe of the liver, diaphragm and oesophagus lie: _____.

    a. Anteriorly   **b.** Posteriorly   **c.** Superiorly   **d.** Inferiorly.

36. The stomach is adapted to allow stretching when filled. The folds that permit this are: _____.

    a. Goblet cells   **b.** Gastric glands   **c.** Villi   **d.** Rugae.

37. Which of the following is not a function of the stomach? _____.

    a. Non-specific defence against microbes   **c.** Secretion of intrinsic factor
    b. Chemical digestion of carbohydrates   **d.** Secretion of gastrin.

38. The flow of gastric juice can be stimulated by the sight or thought of food. This is typical of which phase(s) of gastric acid secretion? _____.

    a. Cephalic phase   **b.** Gastric phase   **c.** Intestinal phase   **d.** All of the above.

39. The gastric phase of gastric secretion is stimulated by secretion of the hormone: _____.

    a. Gastrin   **b.** Cholecystokinin   **c.** Secretin   **d.** Trypsinogen.

? Pot luck

40. The following statements describe the functions of the stomach, but *four* of them contain an incorrect word or phrase. Identify the incorrect statements and give the correct version in the space provided.

    a. The stomach acts as a temporary storage area for foodstuffs, allowing the digestive enzymes time to act.

    _____

    b. Chemical digestion in the stomach includes the action of pepsinogen, an enzyme that acts on proteins and breaks them down to smaller polypeptides.

    _____

    c. The muscular layer of the stomach is essential for mechanical digestion; foodstuffs are churned into a smooth liquid called chyme.

    _____

    d. The stomach has no absorptive function; its environment is too acidic, and absorption cannot occur until the food has been neutralized in the intestines.

    _____

    e. Vomiting, a forceful expulsion of the stomach contents out of the mouth, may be a response to ingestion of irritants or contaminated foods.

    _____

    f. Absorption of iron takes place here; the acid environment of the stomach solubilizes iron salts, an essential step in iron absorption.

    _____

    g. Intrinsic factor is produced here, which is required for absorption of vitamin B_{12} in the ileum.

    _____

    h. The stomach regulates flow of liquidized food into the next part of the digestive tract, the duodenum, through the cricopharyngeal sphincter.

    _____

SMALL INTESTINE

 Matching

41. The small intestine is divided into three sections: the duodenum, the jejunum and the ileum. For each of the statements in Table 12.2, decide to which section it applies by ticking the relevant box in the table.

	Duodenum	Jejunum	Ileum
Longest portion of the small intestine			
Curves around the head of the pancreas			
Vitamin B$_{12}$ is absorbed here			
About 25 cm long			
Middle section			
Ends at the ileocaecal valve			
Flow in is regulated by the pyloric sphincter			
Most digestion takes place here			
About 2 m long			
Flow from here enters the large intestine			
Bile passes into this section			
The pancreas passes its secretions into this section			
Villi present here			
Most absorption takes place here			

Table 12.2 Characteristics of the duodenum, jejunum and ileum

? **Pot luck**

42. Briefly describe the function of an enzyme _____

_____

_____ .

43. Outline the process of peristalsis _____

_____

_____

_____ .

Labelling and colouring

44. Figure 12.10 shows a single villus, only one of the millions that line the small intestine, giving it a velvety appearance. Of the four layers of the wall of the tract, which one forms the villi?

_____

45. Label the main structures shown on Figure 12.10. Colour the arterial blood supply red, the venous drainage blue and the lymphatic vessels green.

46. What is absorbed into the capillaries of the villus?

_____

47. What is absorbed into the central vessel of the villus?

_____

48. What is the name of the large collections of lymphoid tissue found in the intestine?

_____

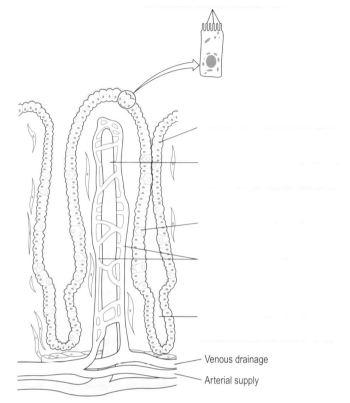

Venous drainage

Arterial supply

Figure 12.10 Highly magnified view of single villus

 Completion

49. The following passage describes chemical digestion in the small intestine. Complete it by scoring out the incorrect option(s) in bold, thus leaving the right one(s).

On a daily basis, the intestine secretes about **1500 ml/2000 ml/2500 ml** of intestinal juices, and its contents are usually acidic, because the contents coming from the stomach are **acidic/between 7.8 and 8.0/very alkaline, to neutralize stomach acid**. In the small intestine, chemical digestion is completed and the end products are absorbed. The main enzyme secreted by the enterocytes is enterokinase, which **breaks down proteins to polypeptides/activates enzymes from the pancreas/neutralizes stomach acid and stops the action of pepsin**. However, other enzymes from accessory structures are passed into the **duodenum/jejunum/ileum** as well.

The pancreas secretes **sucrase/amylase/maltase**, which is important in reducing large sugar molecules to **amino acids/glucose/disaccharides**. In addition, pancreatic lipase breaks down fats into **fatty acids and glucose/amino acids and glycerol/fatty acids and glycerol**, which can be absorbed in the intestine. The third major nutrient group, the proteins, are broken down to **amino acids/dipeptides/polypeptides** by pancreatic **trypsin and chymotrypsin/pepsin and trypsin/chymotrypsin and pepsin**. Pancreatic juice is also rich in **chloride/hydrogen/bicarbonate** ions, important in neutralizing the acid chyme from the stomach.

Bile is made in the **gall bladder/liver/duodenum**, stored in the **gall bladder/liver/duodenum**, and enters the intestine via the **cardiac sphincter/hepatopancreatic sphincter/biliary sphincter**. It has a role to play in fat digestion by breaking fats into **fatty acids and glycerol/tiny droplets/soluble ions**. This increases the action of lipases on the fat.

Even after the multiple digestive actions of these enzymes, the digested proteins and carbohydrates are still not in a readily absorbable form, and digestion is completed by enzymes made by the **enterocytes/goblet cells/lacteals**. Thus, the final stage of protein digestion produces **glucose/amino acids/dipeptides** and the final stage of carbohydrate digestion produces **monosaccharides/glycogen/sucrose**.

MCQs

50. Bile is needed for absorption of which vitamins below (choose all that apply) _____.

    a. A   **b.** B   **c.** C   **d.** D.

51. Pancreatic juice contains which of the following enzymes (choose all that apply) _____.

    a. Peptidase   **b.** Amylase   **c.** Lipase   **d.** Trypsin.

52. In the small intestine, the action of which enzyme(s) breaks down digestible carbohydrates (choose all that apply)? _____.

    a. Peptidase   **b.** Amylase   **c.** Lipase   **d.** Trypsin.

53. Which of the following statements are true (choose all that apply)? _____.

    a. Secretin and cholecystokinin are enzymes.
    b. Secretin and cholecystokinin break fats down into fatty acids and glycerol.
    c. Secretion of pancreatic juice is stimulated by secretion of secretin and cholecystokinin.
    d. Secretion of bile is stimulated by secretion of secretin and cholecystokinin.

54. The final breakdown products of carbohydrates include: _____.

    a. Sucrase   **b.** Sucrose   **c.** Glucose   **d.** Lactose.

Labelling and completion

55. Figure 12.11 shows two intestinal villi that will absorb an assortment of the main nutrients.
    a. Label the parts shown.
    b. The three main nutrients – glucose, amino acids and fatty acids – are represented by different symbols on the figure. Using the distribution of the symbols as a guide, identify which nutrient is represented by:

    ○

    ●

    ☆ _____

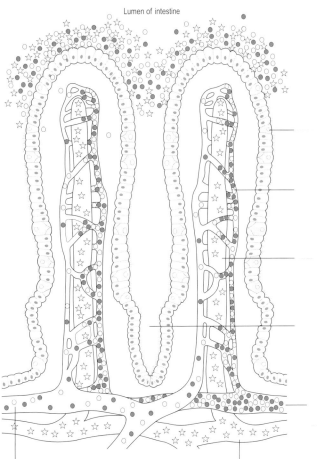

Figure 12.11 The absorption of nutrients

Pot luck

56. Identify whether each vitamin below is absorbed into the lacteals or capillaries of villi?

A _____   B _____

C _____   D _____

E _____   K _____.

57. Some absorbed nutrients pass into the villus simply because there is more of them in the intestine than in the blood – this is simple diffusion. What is the other mechanism by which nutrients can be absorbed?

_____

58. Name three examples of molecules transported by the mechanism you have identified in question 57.

• _____   • _____   • _____.

59. What structures protect the small intestine from infection?

_____

60. Which division of the autonomic nervous system stimulates peristalsis?

_____

61. Which division of the autonomic nervous system stimulates secretion of intestinal juice?

_____

 Completion

62. A huge volume of fluid is secreted into the gastrointestinal tract daily. Given an average daily fluid intake of 1200 ml, complete Figure 12.12, which summarizes the average volumes of fluid secreted, absorbed and eliminated in 24 hours.

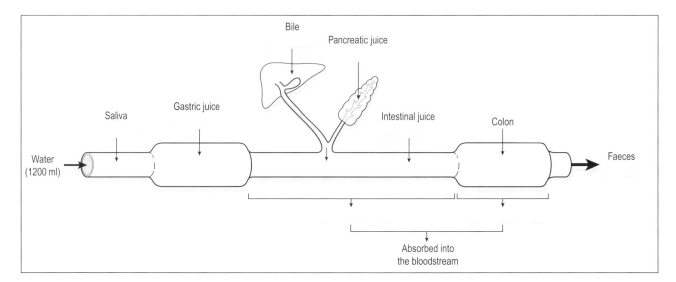

Figure 12.12 Average daily volumes of fluid in the gastrointestinal tract

LARGE INTESTINE, RECTUM AND ANAL CANAL

? MCQs

63. The contents of the large intestine are moved along by: (choose all that apply) _____.

   a. Mass movement   **b.** The gastrocolic reflex   **c.** The ileocaecal reflex   **d.** Defaecation.

64. The transverse colon connects the: _____.

   a. Sigmoid colon and ascending colon   **c.** Ascending colon and descending colon
   b. Rectum and descending colon   **d.** Sigmoid colon and descending colon.

65. Which of the following statements concerning faeces is true? _____.

   a. The bacteria they contain are always harmful
   b. The bacteria they contain are always harmless
   c. They consist mainly of fibrous indigestible material
   d. Their composition is the same as the contents leaving the ileum.

66. Constipation may arise due to: (choose all that apply) _____.
   a. Fast transit along the alimentary canal
   b. Slow transit along the alimentary canal
   c. Presence of commensal microbes in the large intestine
   d. Postponing the need to defaecate.

67. Defaecation: _____.
   a. Is assisted by relaxation of the abdominal muscles
   b. Occurs by reflex during diarrhoea
   c. Occurs by reflex in adults
   d. Occurs by reflex in children.

PANCREAS

 ### Labelling and colouring

68. The pancreas and its associated structures are shown in Figure 12.13. Label and colour the structures indicated.

69. Name the secretions that enter the duodenum at the duodenal papilla.

_____.

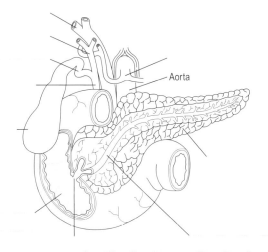

Aorta

Figure 12.13 The pancreas in relation to the duodenum and biliary tract

Matching

70. The pancreas secretes two types of substances, and is considered both an exocrine and an endocrine gland. Complete Table 12.3 using the choices listed and summarize the functions of the pancreas.

Key Choices:
Secretions leave via the pancreatic duct
Control of blood sugar levels
Synthesis takes place in the pancreatic islets
Secretion of glucagon
Substances are passed directly into blood
Role is in digestion

Synthesis takes place in pancreatic alveoli
Secretion of enzymes
Secretion of hormones
Secretions include amylase, lipase and proteases
Passes secretions into duodenum
Secretion of insulin

Exocrine functions	Endocrine functions

Table 12.3 Functions of the pancreas

THE LIVER AND THE BILIARY TRACT

  Colouring and matching

71. Figure 12.14 shows the anterior surface of the liver. Colour and match the following structures:

○ Right lobe
○ Left lobe
○ Gall bladder
○ Falciform ligament
○ Interior vena cava

72. What structure lies immediately above the liver?

_____.

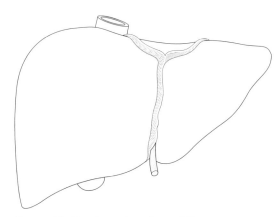

Figure 12.14 Anterior view of liver

Labelling

73. Figure 12.15 shows a magnified transverse section of a liver lobule. Label the structures indicated.

74. In Figure 12.15, blood flows from the small peripheral vessels (shown in groups at the corners of the lobule and associated with a bile duct) towards the central vein in the middle of the lobule. There are two vessels, an artery and a vein. What is the significance of the fact that there is both an arterial and a venous supply to the liver?

_____

_____

_____.

Figure 12.15 A magnified transverse section of a liver lobule

 Completion

75. The following paragraphs describe the functions of the liver. Fill in the blanks.

The liver is involved in the metabolism of carbohydrates; it converts glucose to _____ for storage; the hormone that is important for this is _____. In the opposite reaction, glucose is released to meet the body's energy needs and the important hormone for this is _____. This action of the liver maintains the blood sugar levels within close limits. Other metabolic processes include the formation of waste, including _____, from the breakdown of protein, and _____ from the breakdown of nucleic acids. Transamination is the process by which _____ are made from _____. Proteins are also made here; two important groups of proteins, found in the blood, are the _____ and the _____.

The liver detoxifies many ingested chemicals, including _____ and _____. It also breaks down some of the body's own products, such as _____. Red blood cells and other cellular material such as microbes are broken down in the _____ cells. It synthesizes vitamin _____ from _____, a provitamin found in plants such as carrots, and stores it, along with other vitamins. The liver is also the main storage site of _____ (essential for haemoglobin synthesis).

The liver makes _____, which is stored in the gall bladder and important in digestion of _____. Bile salts are important for _____ in the small intestine, and are themselves reabsorbed there and returned to the liver in the _____. This is called the _____ circulation, and helps to conserve the body's store of bile salts. Bilirubin is released when _____ are broken down (this occurs mainly in the _____ and the _____). Bilirubin is not very soluble so, to increase its water solubility so that it can be excreted in the bile, it is conjugated with _____. On its passage through the intestine, it is converted by bacteria to _____, which is excreted in the faeces; some is, however, reabsorbed and excreted in the urine as _____. If levels of bilirubin in the blood are high, its yellow colour is seen in the tissues as _____.

Labelling, matching and colouring

76. Colour and match the following on Figure 12.16.

- ○ Gall bladder
- ○ Duodenum
- ○ Pancreas

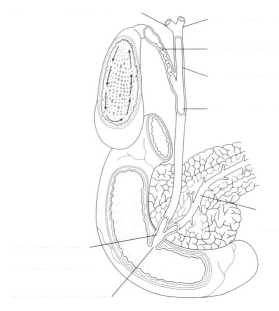

77. Label the structures indicated on Figure 12.16.

78. Insert arrows in one colour to show the direction of flow of bile from the liver into the gall bladder for storage, and in another colour to show the direction of flow from the gall bladder into the duodenum.

- ⇨ – Flow from liver to gall bladder
- ⇨ – Flow from gall bladder to duodenum

Figure 12.16 Flow of bile from liver to duodenum

 MCQs

79. The tiny porous blood vessels that allow easy passage of substances between the blood and hepatocytes are: _____.

  a. Venules   **b.** Arterioles   **c.** Sinusoids   **d.** Capillaries.

80. The vessels within liver lobules that carry bile are the: _____.

  a. Canaliculi   **b.** Capillaries   **c.** Lymphatics   **d.** Sinusoids.

81. Which of the following statements about the liver is true? _____.

  a. The liver has no involvement in metabolism of fats
  b. Regulation of liver function is entirely by the nervous system
  c. Glucose is stored by the liver
  d. Bile is secreted by the liver.

82. The liver receives oxygenated blood via the: _____.

  a. Portal vein   **b.** Hepatic vein   **c.** Hepatic artery   **d.** Renal vein.

83. Nutrients absorbed from the small intestine are carried to the liver via the: _____.

  a. Portal vein   **b.** Hepatic vein   **c.** Hepatic artery   **d.** Renal vein.

84. Bilirubin is a constituent of bile that is formed during: _____.

  a. Breakdown of alcohol (ethanol)   **b.** Haemolysis (breakdown of red blood cells)
  c. Catabolism of glycerol   **d.** Deamination of amino acids.

METABOLISM

 Definitions

Define the following terms:

85. Catabolism _____

_____ .

86. Anabolism _____

_____ .

87. Explain the difference between a kilocalorie and a kilojoule.

_____ .

_____ .

? **MCQs**

88. The end result of energy producing metabolic pathways is the production of which high-energy molecule? _____.

   a. Glucose   **b.** Glycogen   **c.** Citric acid   **d.** Adenosine triphosphate.

89. Which of the following does not refer to metabolism of proteins in the liver? _____.

   a. Deamination   **b.** Detoxification of alcohol   **c.** Transamination   **d.** Synthesis of albumin.

90. Which of the following are not carbohydrates? _____.

   a. Glucose   **b.** Glycogen   **c.** Starch   **d.** Glucagon.

91. The hormone responsible for converting glycogen to glucose is: _____.

   a. Glucagon   **b.** Erythropoietin   **c.** ADH   **d.** Insulin.

92. The hormone responsible for converting glucose to glycogen is: _____.

   a. Glucagon   **b.** Erythropoietin   **c.** ADH   **d.** Insulin.

93. Which of the following does not affect basal metabolic rate (BMR)? _____.

   a. Ingestion of food and starvation   **b.** Age and gender
   c. Resting in a warm room in the postabsorptive state   **d.** Fever.

Labelling, matching and colouring

94. Figure 12.17 shows the biochemical fate of glucose in the cell both in the presence and absence of oxygen. Colour, match and label the arrows to show the three main pathways.

- ○ Glycolysis
- ○ Citric acid (Krebs) cycle
- ○ Oxidative phosphorylation

95. Complete the pathways by inserting the correct metabolic intermediates and products in the spaces provided.

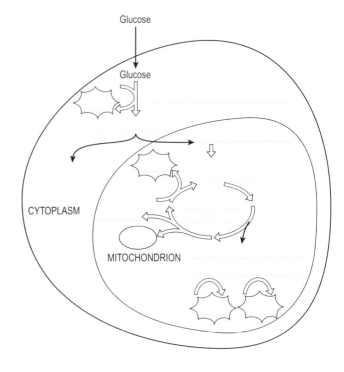

Figure 12.17 Oxidation of glucose

  Matching and colouring

96. Figure 12.18 summarizes the biochemical fates of the main energy sources in the central metabolic pathways, but only the pathways for glucose are complete. Using the labels given below, and by inserting arrows appropriately to indicate conversion of one substance to another, show how proteins and fats also contribute to energy production.

ADP × 4	H$_2$O
Amino acids	Acetyl coenzyme A
ATP × 4	CO$_2$
Ketone bodies	Pyruvic acid
Fatty acids	Oxaloacetic acid
Glycerol	

97. Colour the ATP produced in yellow and the metabolic water in blue.

98. Which substance is essential for the citric acid cycle and oxidative phosphorylation, but is unnecessary for glycolysis?

_____.

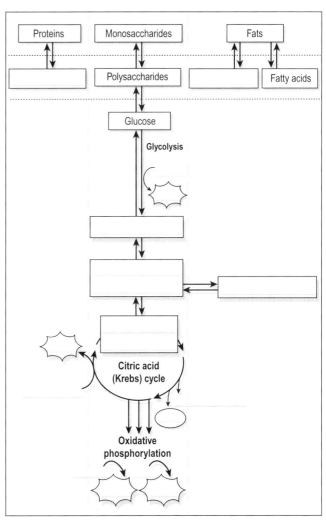

Figure 12.18 Summary of the central metabolic pathways

The urinary system

The urinary system is an important excretory system that plays a vital part in maintaining homeostasis of water and electrolyte concentrations in the body. This chapter will help you to understand how this occurs.

KIDNEYS

  Colouring and labelling

1. Colour and label the structures identified on Figure 13.1.

2. Draw in the left and right adrenal glands on Figure 13.1.

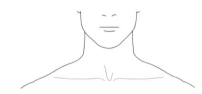

Inferior vena cava

Aorta

Figure 13.1 Parts of the urinary system and some associated structures

 Matching and labelling

3. Match and label the following structures with the parts of the kidney shown on Figure 13.2:

Cortex
Ureter
Medulla
Papilla
Major calyx
Minor calyces
Pelvis
Capsule

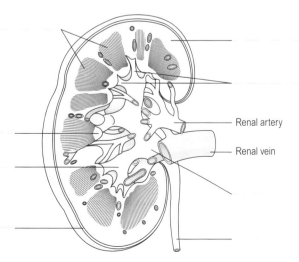

Renal artery

Renal vein

Figure 13.2 A longitudinal section of the kidney

Labelling

4. Label the structures identified on Figure 13.3.

  Colouring and matching

5. Colour and match the following areas:

   ○ Renal medulla
   ○ Renal cortex

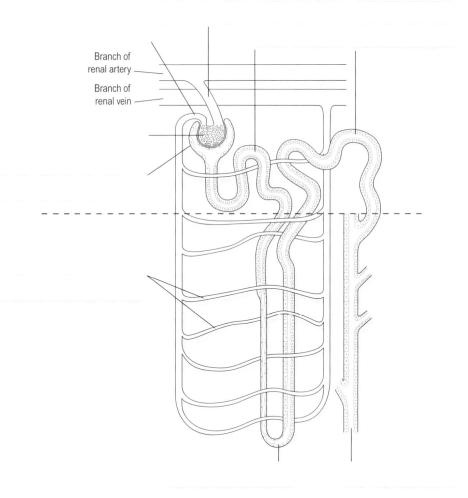

Branch of renal artery

Branch of renal vein

Figure 13.3 A nephron and associated blood vessels

 Completion

6. List the structures that transport urine between the collecting ducts and the ureters.

   Collecting ducts

   _____

   _____

   _____

   _____

   Ureter.

Matching and labelling

7. Match and label the blood vessels from the following list with those indicated on Figure 13.4:

> Renal vein
> Renal artery
> Efferent arteriole
> Afferent arteriole
> Glomerular capillaries
> Capillary network supplying the nephron

8. Between the glomerular capillaries is connective tissue that contains phagocytes, which are also known as

_____.

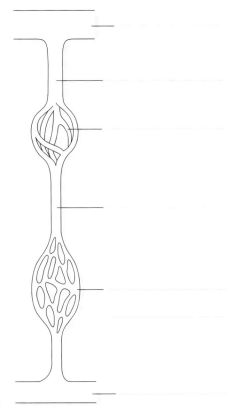

Figure 13.4   The series of blood vessels in the kidney

 MCQs

9. The main structures enter and leave the kidney at the: _____.

   a. Cortex   **b.** Medulla   **c.** Capsule   **d.** Hilum.

10. In the nephron filtration takes place in the: _____.

   a. Medullary loop   **b.** Proximal convoluted tubule   **c.** Glomerulus   **d.** Distal convoluted tubule.

11. Of the 180 litres of filtrate formed daily, how much is reabsorbed? _____.

   a. More than 99%   **b.** 50%   **c.** 10%   **d.** Less than 1%.

12. Which option contains only normal constituents of filtrate? _____.

   a. Water, urea, glucose, creatinine, amino acids   **c.** Water, urea, plasma proteins, glucose, uric acid
   b. Water, urea, leukocytes, mineral salts, amino acids   **d.** Water, erythrocytes, glucose, uric acid, mineral salts.

13. The kidneys are important in regulating homeostasis of: _____.

   a. Water balance   **b.** Electrolyte balance   **c.** pH   **d.** All of the above.

14. Aldosterone is secreted by the adrenal cortex in response to stimulation by: _____.

   a. Angiotensin converting enzyme   **b.** Angiotensin 1   **c.** Angiotensin 2   **d.** Renin.

15. Which hormone is not involved principally in sodium and water balance? _____.

   a. Calcitonin   **b.** Aldosterone   **c.** Atrial natriuretic peptide   **d.** Antidiuretic hormone.

Colouring and completion

16. List the three processes involved in the formation of urine:

- _____

- _____

- _____ .

17. Colour the blood vessels on Figure 13.5.

18. For each process above draw an arrow on Figure 13.5 that indicates:

- the direction of net movement
- the region of the nephron where it occurs.

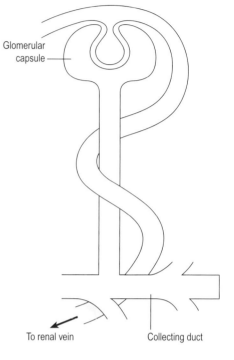

Glomerular capsule

To renal vein   Collecting duct

Figure 13.5 Summary of the processes that form urine

 ## Completion

19. Complete Table 13.1 by identifying the characteristics of normal urine.

Colour	
Specific gravity	
pH	
Average daily volume	

Table 13.1 Characteristics of normal urine

20. Complete Table 13.2 by identifying which of the constituents of blood normally enter the glomerular filtrate and urine (insert 'normal' or 'abnormal' in each box).

Constituent of blood	Presence in glomerular filtrate	Presence in urine
Water		
Sodium		
Potassium		
Glucose		
Urea		
Creatinine		
Proteins		
Uric acid		
Red blood cells		
White blood cells		
Platelets		

Table 13.2 Normal constituents of glomerular filtrate and urine

21. Complete the blanks in the paragraph below to explain the control of water volume in the body.

Water is excreted through the lungs in _____, through the skin as _____ and via the

kidneys as the main constituent of _____. Of these three, the most important in controlling fluid balance

are the _____. The minimum urinary output required to excrete the body's waste products is about

_____ per day. The volume in excess of this is controlled mainly by the hormone _____.

Sensory nerve cells, called _____, detect changes in the osmotic pressure of the blood. They are

situated in the _____. When the osmotic pressure increases, secretion of ADH is _____

and water is _____ by the distal collecting tubules and collecting ducts. These actions result in the osmotic

pressure of the blood being _____. This control system maintains osmotic pressure of the blood within a

narrow range and is known as a _____ system.

Matching

22. Figure 13.6 summarizes the main processes involved in the renin-angiotensin-aldosterone system. Enter the appropriate letter from Figure 13.6 next to each of the key choices listed:

Key choices:

Sodium _____

Potassium _____

Water _____

Increased _____

Volume _____

Vasoconstriction _____

Renin _____

ACE (angiotensin converting enzyme) _____

Aldosterone _____.

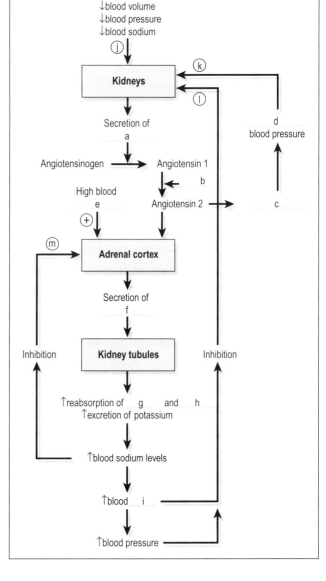

Figure 13.6 Negative feedback regulation of aldosterone secretion

23. Add + or – beside the circles labelled j, k, l and m in Figure 13.6 to indicate whether the arrows stimulate or inhibit the feedback system.

24. Name two other hormones involved in salt and water balance _____

_____.

 Completion

25. Complete the second column of Table 13.3 by adding the site of production of each substance.

Substance	Site of production
Antidiuretic hormone	
Aldosterone	
Angiotensin converting enzyme	
Renin	
Angiotensinogen	
Atrial natriuretic peptide	

Table 13.3 The sites of production of substance that influence the composition of urine

URETERS

 Completion

26. At the upper end, the ureter is continuous with the _____ of the kidney. It passes downwards through

the _____ cavity, behind the _____ in front of the psoas muscle into the _____ cavity

and enters the _____. Urine is propelled down the ureter by the process of _____.

Organs that lie behind the peritoneum can also be described as _____.

 MCQs

27. The ureters are lined with: _____.

    a. Transitional epithelium   **c.** Ciliated columnar epithelium
    b. Stratified squamous epithelium   **d.** Pseudostratified epithelium.

28. The ureters carry urine from: _____.

    a. The nephron to the bladder   **c.** The renal pelvis to the bladder
    b. The renal calyces to the bladder   **d.** The bladder to the external urethral sphincter.

29. Peristalsis in the ureters: _____.

    a. Is an intrinsic property of smooth muscle there   **c.** Is under control of the autonomic nervous system
    b. Is under voluntary control   **d.** Is controlled by hormones.

30. The functions of the ureters include: _____.

    a. Concentration of urine   **b.** Secretion of urine   **c.** Conveying urine   **d.** Storage of urine.

URINARY BLADDER

Colouring and matching

31. Colour and match the following layers of the bladder with those shown on Figure 13.7:

○ Fibrous layer
○ Smooth muscle layer
○ Inner layer

32. Outline the physiological significance of the angle at which the ureter enters the bladder.

_____

_____.

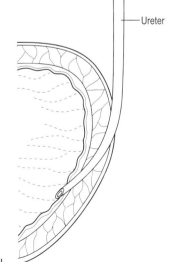

Ureter

Figure 13.7 The position of the ureter where it passes through the bladder wall

Colouring, matching and labelling

33. Colour and match the following structures on Figures 13.8A and B:

○ Rectum
○ Pubic bone
○ Anterior abdominal wall
○ Bladder
○ Right ureter

34. Colour and label the remaining structures on Figures 13.8A and B.

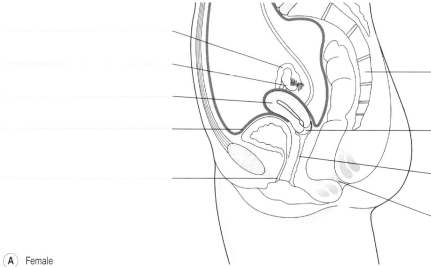

(A) Female

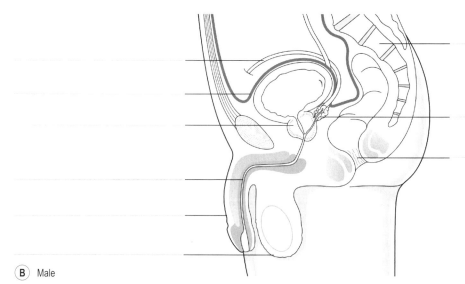

(B) Male

Figure 13.8 Organs associated with the bladder in the male and female

Completion

35. Fill in the blanks in the paragraph below describing the structure of the bladder.

The bladder acts as a _____ for urine. When empty, its shape resembles a _____ and it becomes more

_____ as it fills. The posterior surface is the _____ and the bladder opens into the urethra at its lowest point,

the _____. The bladder wall is composed of three layers. The outer layer is composed of _____

and contains _____ and _____ vessels. The muscular layer is formed by _____ muscle

arranged in _____ layers. Collectively this is called the _____ and when it contracts the bladder

_____. The inner layer is the _____ and it is lined with _____. Three orifices on the

posterior bladder wall form the_____. The two upper openings are formed when each _____ enters

the bladder and the lower one is the opening of the _____.

? **MCQs**

36. The functions of the bladder include: _____.

   a. Concentration of urine   **b.** Manufacture of urine   **c.** Secretion of urine   **d.** Storage of urine.

37. The bladder: _____.

   a. Is completely covered by peritoneum   **c.** Contents are normally sterile (free of microbes)
   b. Expels urine by reflex action in adults   **d.** Is situated in the abdominal cavity.

38. The muscle layer in the bladder is also known as: _____.

   a. The trigone   **b.** The detrusor   **c.** The submucosa   **d.** The internal sphincter.

URETHRA

? **MCQs**

39. Which layer of the urethral wall consists of stratified squamous epithelium at the distal end? _____.

   a. Mucosa   **b.** Submucosa   **c.** Muscle   **d.** Trigone.

40. How many layers of tissue are found in the wall of the urethra?_____.

   a. One   **b.** Two   **c.** Three   **d.** Four.

41. The internal urethral sphincter is composed of elastic tissue and: _____.

   a. Fibrous tissue   **b.** Smooth muscle   **c.** Skeletal muscle   **d.** Cardiac muscle.

42. The external urethral sphincter is composed of: _____.

   a. Fibrous tissue   **b.** Smooth muscle   **c.** Skeletal muscle   **d.** Cardiac muscle.

MICTURITION

 ### Matching

43. Select key choices from the list below to complete the blank spaces in the paragraph to describe the differences in micturition in infants and adults.

Key choices	
Brain	Over-ridden
Contraction	Relaxation
Detrusor	Spinal reflex
External	Stretching
Internal	Voluntary

As the bladder fills and becomes distended, receptors in the wall are stimulated by _____. In infants this

initiates a _____ and micturition occurs as nerve impulses to the bladder cause _____ of the

_____ muscle and _____ of the _____ urethral sphincter. When the nervous system is fully

developed the micturition reflex is stimulated but sensory impulses pass upwards to the _____. By conscious

effort, the reflex can be _____. In addition to the processes involved in infants, there is _____

relaxation of the _____ urethral sphincter.

Definitions

Define the following terms:

44. Polyuria _____ .

45. Glycosuria _____ .

46. Polydipsia _____ .

47. Ketonuria _____ .

48. Haematuria _____ .

49. Proteinuria _____ .

Applying what you know

50. In some kidney disorders the glomerular capillaries become more permeable. Substances that do not normally cross may then enter the filtrate and are excreted in the urine. Give three such examples.

    • _____

    • _____

    • _____.

51. In diabetes mellitus raised blood sugar levels result in high concentrations of glucose in the glomerular filtrate. Explain the effects of this.

    _____

    _____

    _____.

52. In diabetes insipidus secretion of ADH is impaired. Describe the effects of this on urine output.

    _____.

53. ACE inhibitors are a group of drugs that inhibit the action of angiotensin converting enzyme. State their effects on blood pressure.

    _____.

54. When a kidney stone becomes lodged in a ureter it causes acute pain:

    a. Explain why this occurs. _____

    _____

    b. What is the medical term for this condition? _____

    c. As urine accumulates above the blockage, what consequences may arise? _____

    _____.

    d. What is the medical term for kidney stones? _____

    e. Find out which painless treatment may be used to 'shatter' small kidney stones _____.

55. Cystitis is inflammation of the bladder usually caused by entry of commensal bacteria from the bowel, e.g. *Escherichia coli*. Explain why women are anatomically more prone to this condition than men.

    _____

    _____.

The skin

The skin completely covers the body and is continuous with the membranes that line the body orifices. This chapter will help you to learn about its structure and functions.

STRUCTURE OF THE SKIN

  Colouring and labelling

1. Colour the capillaries on Figure 14.1.

2. Colour and label the structures identified on Figure 14.1.

3. Colour the following layers of the skin:

| ○ Epidermis | ○ Dermis | ○ Subcutaneous tissue |

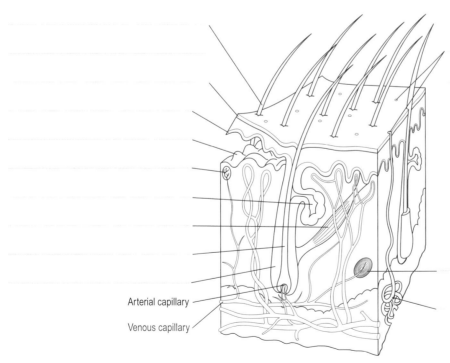

Arterial capillary

Venous capillary

Figure 14.1 The main structures in the skin

? MCQs

4. The epidermis: _____.

    a. Contains Pacinian corpuscles, receptors that are sensitive to deep pressure
    b. Lies directly above adipose tissue
    c. Consists of several layers of flattened epithelial cells
    d. Is supplied with blood by an extensive capillary network.

5. Complete regeneration of the epidermis takes about: _____.

    a. One week   **b.** One month   **c.** Two months   **d.** Three months.

6. Waterproofing of the skin is provided by: _____.

    a. Keratin   **b.** Melanin   **c.** Tyrosine   **d.** Sweat.

7. The colour of the skin may be affected by (choose all that apply): _____.

    a. Presence of high levels of bile pigments in the blood
    b. Low saturation of blood haemoglobin
    c. The amount of melanin
    d. Keratinization of epidermal cells.

8. Which of the statements are true (choose all that apply)? _____.

    a. The deepest layer of the epidermis is the stratum corneum
    b. The dermis contains keratinized cells
    c. The dermis contains collagen and elastic fibres
    d. The dermis provides the epidermis with nutrients.

9. Sebum is secreted by: _____.

    a. Arrector pili
    b. Sweat glands
    c. Sebaceous glands
    d. Sweat glands and sebaceous glands.

FUNCTIONS OF THE SKIN

 Matching

10. Match the correct key choice with the appropriate statement.

Key choices:		
Langerhans cells	Evaporation	Non-specific defence mechanism
Sensory nerve endings	Convection	Absorption
Conduction	Vasodilation	Vitamin D

    a. Stimulation may initiate a reflex response: _____

    b. Occurs when objects in contact with the skin take up heat: _____

    c. Results in increased blood flow and is recognized by redness of the skin: _____

    d. Formed by conversion of 7-dehydrocholesterol by UV rays in sunlight: _____

    e. Specialized immune cells: _____

    f. Takes place when heat is used to convert water in sweat to water vapour: _____

    g. The mechanism by which a limited number of substances gain entry to the body: _____

    h. Occurs as cool air replaces warmed air which has risen from the body: _____

    i. A means of protection against many different potential dangers: _____.

? Pot luck

11. State whether each process results in heat loss or heat gain.

a. Shivering _____

b. Sweating _____

c. Conduction _____

d. Radiation _____

e. Vasodilation _____

f. Vasoconstriction _____

g. Evaporation _____

h. Convection _____.

12. Some of the statements about the skin are incorrect. Identify which ones and correct the errors.

a. Insensible water loss is around 1 litre per day.

_____

b. The distribution of sensory receptors in the skin is the same in all areas of the body.

_____

c. The skin can absorb some substances.

_____

d. The skin is involved in the formation of vitamin A.

_____

✎ Completion

13. Fill in the blanks to complete the paragraphs describing temperature regulation.

The temperature regulating centre is situated in the _____ and is responsive to the temperature of circulating _____. When body temperature rises, sweat glands are stimulated by the _____. The _____ centre in the medulla oblongata controls the diameter of small arteries and _____ and therefore the amount of _____ circulating in the dermis. When body temperature rises the skin capillaries _____ and extra blood near the surface increases heat loss by _____, _____ and _____. The skin is warm and _____ in colour. When body temperature falls, arteriolar vasoconstriction conserves heat and the skin is _____ and feels cool.

　Fever is often the result of _____. During this process there is release of chemicals, also called _____, from damaged tissue. These chemicals act on the _____ which releases prostaglandins that reset the temperature thermostat to a _____ temperature. The body responds by activating heat promoting mechanisms, e.g. _____ and _____, until the new temperature is reached. When the thermostat is reset to the normal level, heat loss mechanisms are activated. There is vasodilation and profuse _____ until body temperature returns to the normal range again.

WOUND HEALING

 Completion

14. Complete Table 14.1 by identifying factors that affect the rate of wound healing.

	Promote wound healing	Impair wound healing
Systemic factors	_____ _____	_____ _____
Local factors	_____	_____

Table 14.1 Factors affecting the rate of wound healing

 Definitions

Define the following terms:

15. Primary wound healing _____.

16. Secondary wound healing _____.

  Colouring and labelling

17. Identify the stages of wound healing shown in Figure 14.2.

18. Colour the following on Figure 14.2:

○ Fibroblasts
○ Phagocytes

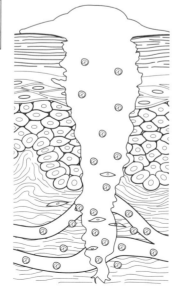 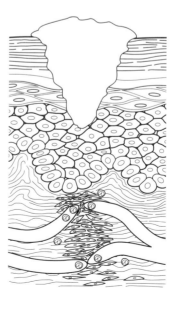 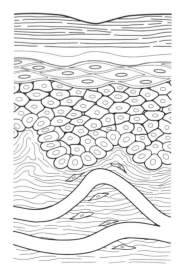

Figure 14.2 Stages in primary wound healing

Matching

19. Match the key choices with the statements. Key choices may be used more than once.

> *Key choices:*
> Phagocytes   Scar tissue   Granulation tissue
> Slough   Fibroblasts

Statements

a. Clotted blood and wound debris are removed by _____

b. Collagen fibres are produced by _____

c. Necrotic tissue may also be called _____

d. These cells travel to a wound in its blood supply _____

e. Tissue that consists of capillary buds, phagocytes and fibroblasts is called _____

f. Tissue formed from fibrous tissue is called _____.

? **MCQs**

20. Fibrosis that accompanies wound healing may result in (choose all that apply): _____.

    a. Its replacement by new tissue the same as that previously in the affected area
    b. Tissue shrinkage
    c. Adhesions
    d. Scar formation.

21. Suppuration is a complication of wound healing which may result in (choose all that apply): _____.

    a. Formation of a haematoma   **b.** Contusions   **c.** Boils   **d.** Abscesses.

22. Suppuration is most commonly caused by: _____.

    a. Viruses   **b.** Bacteria   **c.** Dehydration   **d.** Haemorrhage.

23. Following wound healing, bands of fibrous tissue that extend across a joint limiting its movement are also known as: _____.

    a. Adhesions   **b.** Fistulae   **c.** Contractures   **d.** Fractures.

✎ Applying what you know

24. Define the term hypothermia. _____

_____.

25. Outline why the skin is pale in hypothermia. _____

_____.

26. Explain why shivering starts and stops again as core body temperature decreases.

_____

_____

_____.

Resistance and immunity

CHAPTER 15

From life in the womb to the moment of death, an individual is under constant attack from an enormous range of potentially harmful invaders, including bacteria, viruses, parasites and foreign (non-self) cells. The body has therefore developed a wide range of protective measures, both specific and non-specific, which will be considered in this chapter.

? Pot luck

1. Decide which of the following defence mechanisms is specific (S) and which is non-specific (N/S).

   a. skin _____

   b. phagocytosis _____

   c. antibodies _____

   d. complement _____

   e. lysozyme _____

   f. lymphocyte activation _____

   g. inflammation _____

   h. T-cell production _____

   i. natural killer cells _____

   j. interferons _____

   k. B-cell production _____

   l. stomach acid _____.

NON-SPECIFIC DEFENCE MECHANISMS

? MCQs

2. The system of about 20 defensive proteins found in body fluids and tissues is called: _____.

   a. Antibodies　　　**b.** Interferons　　　**c.** Complement　　　**d.** Lysozyme.

3. The function of interferons is to: _____.

   a. Attach to and destroy antigens
   b. Break down bacteria in phagocytic granules
   c. Prevent the spread of viruses to healthy cells
   d. Attract white blood cells into infected tissues.

4. Histamine: _____.

   a. Increases vascular permeability
   b. Causes vasoconstriction
   c. Is released from damaged cell membranes
   d. Relaxes smooth muscle.

5. Prostaglandins: _____.

   a. Are stored preformed in cell granules with heparin
   b. Reduce the pain of tissue damage
   c. Are only found in platelets
   d. Increase vascular permeability in inflammation.

6. Which of the following is a function of complement? _____.

   a. Production of pain
   b. Increased vascular permeability
   c. Destruction of bacteria
   d. Anticoagulant.

Colouring and labelling

7. Which defensive process is shown in Figure 15.1?

   _____

8. Label the structures shown on Figure 15.1.

9. Briefly describe the events being shown in each part of Figure 15.1.

   A. _____

   B. _____

   C. _____

   D. _____ .

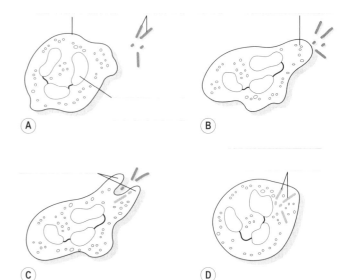

Figure 15.1 Action of neutrophils

 ### Colouring, matching and labelling

10. Figure 15.2 summarizes the main events of the acute inflammatory reaction. Colour, match and identify the cells labelled A, B, C and D and their functions.

    ○ A _____

    _____

    ○ B _____

    _____

    ○ C _____

    _____

    ○ D _____

    _____ .

11. Figure 15.2 shows three different inflammatory mediators. Identify them according to the information given by colouring and matching the symbols in the key.

    ○ Stored in preformed granules by cell C

    _____

    ✳ Causes pain by acting on free nerve endings

    _____

    ⬡ Family of mediators made from cell membranes

    _____ .

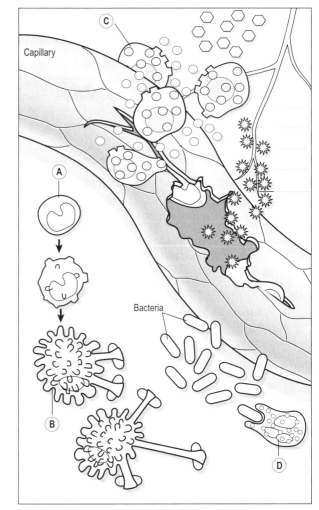

Figure 15.2 The inflammatory response

? Pot luck

12. List the five main signs of the inflammatory response:

- _____

- _____

- _____

- _____

- _____

13. In the following five statements about inflammation, the first statement is always true, but does not necessarily correspond to the reason given. Only one of the reasons corresponds to the initial statement. Identify the four incorrect reasons and correct them in the spaces given below.

Statement		Reason
A. Blood flow to an inflamed area increases	because	elevated tissue temperatures cause vasodilation.
B. An inflamed area swells	because	hydrostatic pressure at the arterial ends of local blood vessels falls.
C. White blood cells migrate into inflamed tissues	because	they are attracted by histamine.
D. The increased tissue temperature in inflammation is beneficial	because	it promotes the activity of phagocytes.
E. The movement of fibrinogen from the bloodstream into inflamed and damaged tissues is beneficial	because	it reduces blood loss from damaged tissue.

A _____

B _____

C _____

D _____

E _____.

Completion

14. Table 15.1 lists some important information about significant inflammatory mediators. Complete the table by filling in the blanks.

Substance	Made by	Trigger for release	Main actions
	Mast cells in tissues and basophils in blood; stored in cytoplasmic granules		
Serotonin (5-HT)			
Prostaglandins			
	Liver; also mast cells, where it is stored in granules with histamine		
			Pain, by acting directly on free nerve endings, vasodilation

Table 15.1 Summary of some important inflammatory mediators

 Matching

15. Match the features of the inflammatory response in list A to the items in Table 15.2.

List A

a. Neutrophil recruitment
c. Complement activation
e. Lymphocyte accumulation
g. Tissue swelling
i. Erythema

b. Increased temperature
d. Plasma protein leakage into tissues
f. Pain
h. Vasodilation

Limits movement of the damaged area	
Cushions the area of damaged tissue	
Associated with chronic inflammation	
Brings antibodies into inflamed tissues	
Increases blood supply for tissue repair	
Stimulates chemotaxis	
Responsible for phagocytosis	
Increases the temperature of damaged area	
First protective cell to arrive at area of damaged tissue	
Enhances phagocytosis of bacteria	
Promotes activity of phagocytes	

Table 15.2 Features of the inflammatory response

IMMUNITY

 Completion

16. Table 15.3 lists various characteristics of the two main populations of lymphocyte. Complete the table.

Characteristic	T-lymphocyte	B-lymphocyte
Shape of nucleus		
Site of manufacture		
Site of post-manufacture processing		
Nature of immunity involved		
Specific or non-specific defence		
Production of antibodies		
Processing regulated by thymosin		

Table 15.3   Lymphocyte characteristics

17. Table 15.4 lists the five categories of antibody. Complete the right-hand column with the functions of each type.

Antibody type	Function
IgA	
IgD	
IgE	
IgG	
IgM	

Table 15.4   The five types of antibody and their functions

Matching, labelling and colouring

18. Figure 15.3 shows how the production of the different types of T-lymphocyte comes about. Identify the different types of cell and the function of each using the key below.

   ○ Macrophage

   Function: _____

   _____

   ○ T-lymphocyte, unspecialized

   Function: _____

   _____

   ○ Cytotoxic T-lymphocyte

   Function: _____

   _____

   ○ Helper T-lymphocyte

   Function: _____

   _____

   ○ Memory T-lymphocyte

   Function: _____

   _____

   ○ Suppressor T-lymphocyte

   Function: _____

   _____.

19. Label the remaining items indicated on Figure 15.3.

20. Colour the receptors for antigen on all cells you see.
   What is significant about these receptors when you compare them between cell types?

   _____

   _____

21. The process of proliferation and differentiation shown in the centre of the diagram is also known as what?

   _____

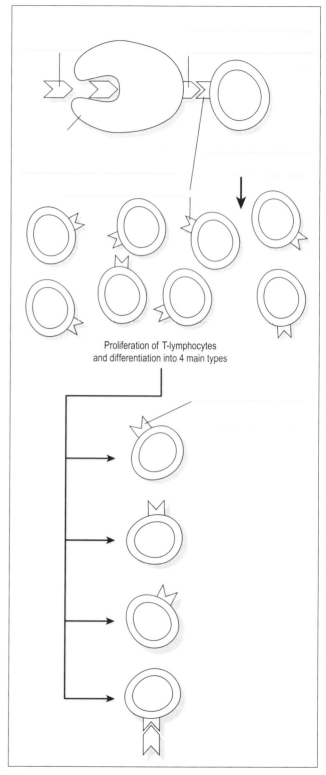

Proliferation of T-lymphocytes and differentiation into 4 main types

Figure 15.3 Production of subtypes of T-lymphocytes

Matching, labelling and colouring

22. Figure 15.4 shows how the production of the different types of B-lymphocyte comes about. Colour and match the different types of cell and the function of each using the key below.

○ Helper T-lymphocyte

Function: _____

○ B-lymphocyte

Function: _____

○ Memory B-lymphocyte

Function: _____

○ Plasma cell

Function: _____.

23. Label the remaining items indicated on Figure 15.4.

24. Explain what is represented at A in Figure 15.4.

_____

_____.

25. Colour the receptors for antigen on any cell you see. What is significant about these receptors when you compare them between cell types?

_____

_____.

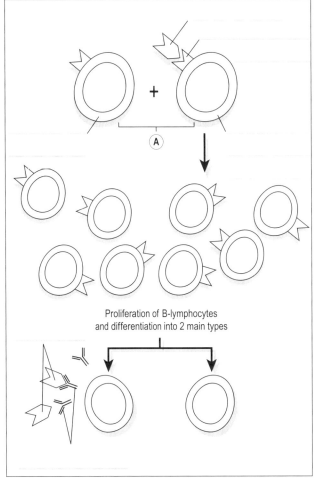

Figure 15.4   Production of subtypes of B-lymphocytes

? | **MCQs**

26. Which cell type, shown in both Figures 15.3 and 15.4, is an important link cell between non-specific and specific defence?_____.

   a. Macrophage   **b.** Cytotoxic T-lymphocyte   **c.** Helper T-lymphocyte   **d.** Memory T-lymphocyte.

27. Antibodies are produced by: _____.

   a. Macrophages   **c.** Memory B-lymphocytes
   b. Helper T-lymphocytes   **d.** Plasma cells.

28. The general function of antibodies include (choose all that apply): _____.

   a. Activating complement   **c.** Neutralizing bacterial toxins
   b. Recruitment of T- and B-lymphocytes   **d.** Stimulating clonal expansion of B-lymphocytes.

29. Plasma cells are formed from: _____.

   a. T-lymphocytes   **c.** Macrophages
   b. B-lymphocytes   **d.** Neutrophils.

30. Which immune cell is responsible for activating the B-lymphocyte response following T-lymphocyte proliferation? _____.

   a. Helper T-lymphocyte   **c.** Memory T-lymphocyte
   b. Cytotoxic T-lymphocyte   **d.** Suppressor T-lymphocyte.

Completion

31. The following paragraph discusses the antibody response to antigen exposure. Complete it by filling in the blanks.

When the body is exposed to an antigen for the first time, the immune response can be measured as antibody

levels in the blood after about _____ weeks; this is the _____ response. Antibody levels fall

thereafter, and do not rise again unless there is a second exposure to the same antigen, which stimulates a

_____ response, which is different from the first in that it is much _____ and antibody levels

become much _____. After having been exposed to an antigen, an individual may develop immunity to

it, provided he has produced a population of _____ cells.

32. Immunity may be acquired in different ways, and the nature of the immunity may vary. Complete Table 15.5 by ticking the appropriate boxes relevant to each of the four listed types of immunity.

Characteristic	Active natural	Active artificial	Passive natural	Passive artificial
An example is a baby's consumption of antibodies in its mother's milk				
Long-lived protection				
Involves production of memory cells				
An example is vaccination				
Short-lived protection				
An example is infusion of antibodies				
Involves production of antibodies by the individual				
An example is a child catching chickenpox at school				
Specific				

Table 15.5 The four types of acquired immunity

The musculoskeletal system

The musculoskeletal system consists of the bones of the skeleton, their joints and the skeletal (voluntary) muscles that move the body. This chapter will help you to understand the structure and function of each component of the musculoskeletal system.

BONE

  Colouring and matching

1. Colour and match the following parts of the long bone on Figure 16.1:

○ Compact bone
○ Spongy bone
○ Medullary canal
○ Nutrient artery
○ Articular cartilage
○ Epiphyseal plate

2. Name the vascular membrane that covers bone:

_____.

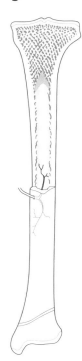

Figure 16.1 A mature long bone – partially sectioned

 Completion

3. Give an example of each type of bone:

   a. Long _____

   b. Short _____

   c. Irregular _____

   d. Flat _____

   e. Sesamoid _____.

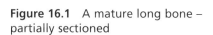 **Labelling**

4. In Figure 16.2, which shows the microscopic structure of compact bone, label the parts indicated.

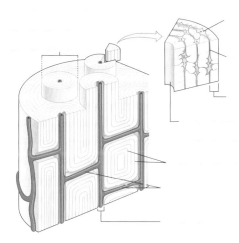

Figure 16.2 Microscopic structure of compact bone

The musculoskeletal system

 Matching

5. Match the key choices with the statements in Table 16.1.

Key choices:
Sesamoid bones
Flat bones
Long bones
Chondroblasts
Lacunae
Osteon
Osteoblasts
Osteoclasts
Osteocytes
Red bone marrow
Trabeculae
Interstitial lamellae
Spongy bone
Spongy bone
Cortical bone

	Looks like a honeycomb to the naked eye
	Cells that lay down bone
	Cells that break down bone
	Haversian system
	Cancellous bone
	Compact bone
	Form the framework of spongy bone
	Cells that lay down cartilage
	Mature osteoblasts
	Remains of old osteons
	Spaces between lamellae that contain osteocytes
	Found mainly in spaces within spongy bone
	Develop from membrane models
	Develop from tendon models
	Develop from cartilage models

Table 16.1 Characteristics of bone

 Completion

6. To make the following paragraph read correctly, delete the incorrect options in bold. Take care, there may be more than one correct answer in each set of options!

Bone tissue develops in the fetus from **adipose/connective/epithelial** tissue models. This process is called

osteogenesis/oogenesis/ossification and is **usually complete/incomplete/always complete** at birth. The main

constituent of bone is **water/calcium salts/bone marrow**, and the organic component is primarily **cartilage/

phosphate/collagen**. During life, bone growth is stimulated by **parathormone/thyroxine/oestrogen**, but its density

is decreased by **calcium/testosterone/lack of exercise**.

Pot luck

7. Name the bony landmarks described below:

    a. A hollow or depression   _____

    b. A ridge of bone separating two surfaces   _____

    c. A large rough bony projection for muscle and ligament attachment   _____

    d. A small hole through a bone   _____

    e. A smooth rounded projection for making a joint   _____

    f. A tube-shaped cavity in bone   _____

    g. An immovable joint between skull bones   _____

    h. A narrow slit in bone   _____

    i. A sharp bony ridge (two terms for this)   _____   _____

    j. A small flat surface for making a joint   _____

    k. A hollow cavity within a bone   _____.

8. Find each answer concealed within the grid. Words run horizontally, vertically and diagonally.

S	L	C	I	G	D	A	W	T	T	R	T
I	N	O	K	H	C	R	E	S	T	E	R
N	E	N	B	Y	U	C	M	P	S	I	O
U	G	D	D	A	A	Y	U	R	O	L	C
S	I	Y	F	F	O	S	S	A	C	Y	H
U	R	L	N	K	I	G	D	A	Y	U	A
B	M	E	A	T	U	S	R	E	O	M	N
O	I	G	D	A	W	T	R	R	O	S	T
R	L	F	I	S	S	U	R	E	I	P	E
D	O	I	G	D	T	A	W	U	R	I	R
E	N	K	F	U	B	Y	V	S	O	N	L
R	G	D	S	A	F	O	R	A	M	E	N

Completion

9. Define the different types of fractures in Table 16.2.

Type of fracture	Characteristics
Simple	
Compound	
Pathological	

Table 16.2 Types of fractures

? MCQs

10. The haematoma that forms around the broken bone and ends at a fracture site is mainly: _____

    a. Clotted blood
    b. New, immature bone

    c. Osteoblasts for repair
    d. Proliferating granulation tissue.

11. Callus is composed of: _____

    a. Cartilage, spongy bone and compact bone
    b. Cartilage and spongy bone

    c. Spongy bone and compact bone
    d. Cartilage and compact bone.

12. Recanalization of new bone is carried out by: _____

    a. Osteoblasts
    b. Macrophages

    c. Osteocytes
    d. Osteoclasts.

13. Which of the following is true? _____.

    a. Macrophages clear the fracture site of debris
    b. Growth of granulation tissue follows initial callus deposition

    c. The broken ends of bone are initially united with compact bone
    d. Fibroblasts develop into new osteoblasts for bone repair.

✎ Applying what you know

14. List four factors that delay healing of fractures:

    • _____

      _____

    • _____

      _____

    • _____

      _____

    • _____

      _____

AXIAL SKELETON

   Colouring, matching and labelling

15. Colour and match the following parts of the skeleton:

- ○ Axial skeleton
- ○ Bones of shoulder girdle and the upper limb
- ○ Bones of pelvic girdle and the lower limb

16. Name the bones identified on Figure 16.3.

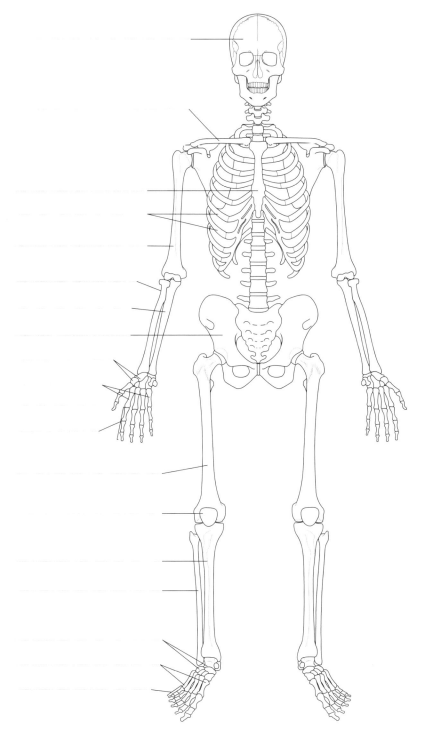

Figure 16.3 The skeleton

Colouring, matching and labelling

17. Colour and match the following skull bones shown on Figure 16.4:

○ Ethmoid bone
○ Fontal bone
○ Lacrimal bone
○ Occipital bone
○ Maxilla
○ Mandible
○ Nasal bone
○ Parietal bone
○ Sphenoid bone
○ Temporal bone
○ Zygomatic bone

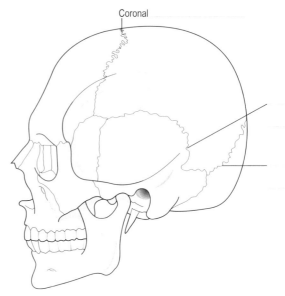

Coronal

Figure 16.4 The bones of the skull and their sutures

18. Label the skull sutures identified on Figure 16.4.

 Completion

19. State which bone(s) of the skull:

a. Form(s) the posterior part of the hard palate: _____

b. Form(s) the anterior part of the hard palate: _____

c. Form(s) the main part of the nasal septum: _____

d. Form(s) the most posterior part: _____

e. Give(s) rise to the mastoid processes: _____

f. Contain(s) the hypophyseal fossa: _____

g. Form(s) the cribriform plate: _____

h. Contain(s) the middle ear: _____

i. Contain(s) the foramen magnum: _____

j. Contain(s) the foramina for the nasolacrimal ducts: _____.

20. Fill in the blanks to complete the description of sinuses and fontanelles of the skull.

Sinuses contain _____ and are found in the _____, _____, _____ and

_____ bones. They all communicate with the _____ and are lined with _____. Their

functions are to give _____ to the voice and _____ the bones of the face and cranium.

    Fontanelles are distinct _____ areas of the skull in infants and are present until _____ is

complete and the skull bones fuse. The largest are the _____ fontanelle, present until _____

months, and the _____ fontanelle that usually closes over by _____ months of age. Their

presence allows for moulding of the baby's _____ during childbirth.

Colouring and matching

21. Colour and match the following parts of Figure 16.5:

- ○ Intervertebral discs
- ○ Cervical vertebrae
- ○ Thoracic vertebrae
- ○ Lumbar vertebrae
- ○ Sacrum
- ○ Coccyx

22. State the number of:

    a. Cervical vertebrae _____

    b. Thoracic vertebrae _____

    c. Lumbar vertebrae _____

    d. Vertebrae that fuse to form the sacrum _____

    e. Vertebrae that fuse to form the coccyx _____.

23. On Figure 16.5, colour and match the open arrows to show the:

- ○ 1st secondary curve
- ○ 2nd secondary curve
- ○ two areas of the primary curve.

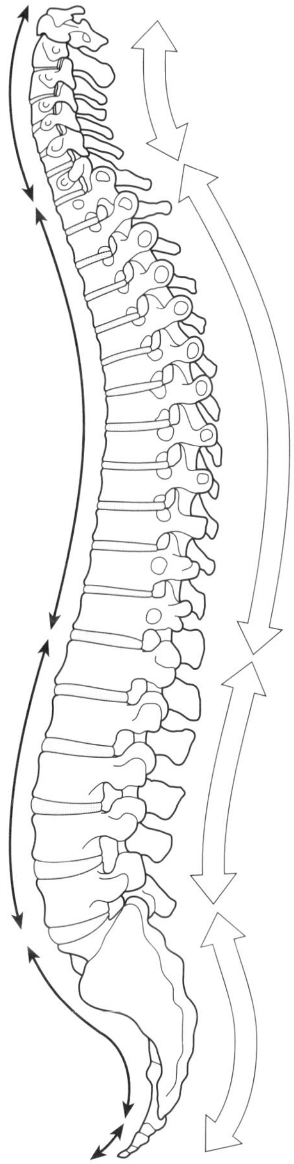

Figure 16.5 The vertebral column. Lateral view

Labelling and matching

24. Match and label the parts of a typical vertebra shown in Figure 16.6:

Body

Lamina

Pedicle

Spinous process

Superior articular process

Transverse process

Vertebral foramen

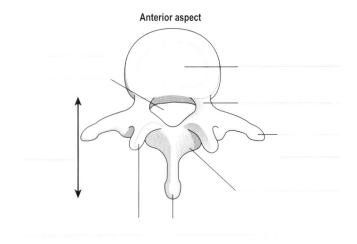

Anterior aspect

Figure 16.6 A lumbar vertebra showing features of a typical vertebra, viewed from above

 Matching

25. Match the key choices listed with the statements about the vertebral column below:

Key choices:
Odontoid process
Intervertebral disc
Sacrum
Axis
Thoracic vertebrae
Vertebral foramen
Nucleus pulposus
Atlas
Coccyx
Annulus fibrosus
Transverse foramen
Intervertebral foramen
Lumbar vertebrae

a. Consists of four fused vertebrae: _____

b. Part of a cervical vertebra containing the vertebral artery: _____

c. Vertebrae that articulate with ribs: _____

d. Space for passage of spinal nerves: _____

e. First cervical vertebra: _____

f. Second cervical vertebra: _____

g. Articulates with the ilium to form the sacroiliac joints _____

h. The largest vertebrae: _____

i. Forms the body of the atlas: _____

j. Part of the vertebrae containing the spinal cord: _____

k. Separates the bodies of adjacent vertebrae: _____

l. The outer part of the intervertebral disc: _____

m. The central core of the intervertebral disc: _____.

Colouring and matching

26. Colour and match the following structures shown in Figure 16.7:

○ Costal cartilages
○ Ribs attached to sternum
○ Floating ribs
○ Sternum
○ Clavicles
○ Vertebrae

27. Name the nerves running in the groove found on the

underside of each rib _____.

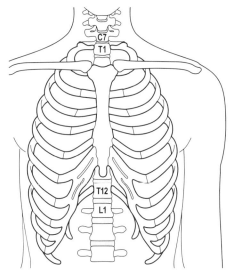

Figure 16.7 The thoracic cage. Anterior view

APPENDICULAR SKELETON

  Matching and labelling

28. Label Figure 16.8 to show the parts of the scapula using the list below.

Parts of the scapula
Acromion process
Coracoid process
Glenoid cavity
Inferior angle
Infraspinous fossa
Lateral border
Medial border
Spine
Supraspinous fossa
Superior angle
Superior border

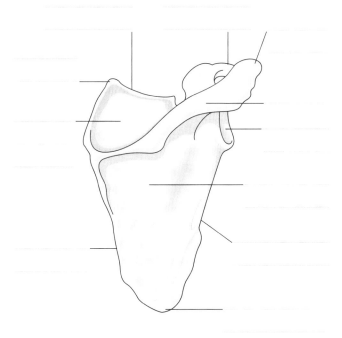

Figure 16.8 The right scapula. Posterior view

Colouring, matching and labelling

29. Match and label the parts of the humerus on Figure 16.9 using the terms listed below.

Parts of the humerus
Head
Neck
Shaft
Greater tubercle
Lesser tubercle
Bicipital groove
Capitulum
Coronoid fossa
Deltoid tuberosity
Lateral supracondylar ridge
Medial supracondylar ridge
Lateral epicondyle
Medial epicondyle
Trochlea

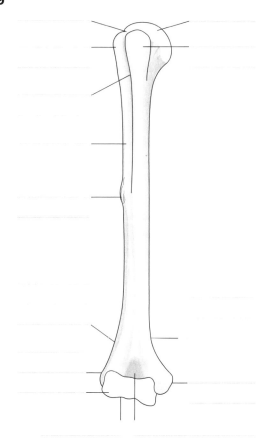

Figure 16.9 The right humerus. Anterior view

30. On Figure 16.10, colour and match the following:

○ Radius

○ Ulna

○ Olecranon process

○ Interosseous membrane

31. Identify and label the structures shown on Figure 16.10.

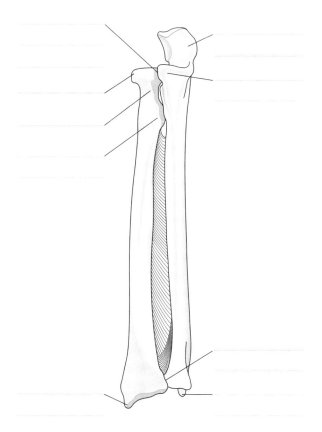

Figure 16.10 The right radius and ulna

32. Colour and match the following parts of Figure 16.11:

- ○ The carpal bones
- ○ The metacarpal bones
- ○ The proximal phalanges
- ○ The middle phalanges
- ○ The distal phalanges

33. Label the bones of the wrist shown on Figure 16.11.

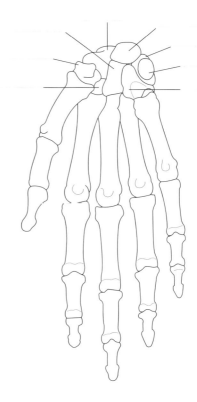

Figure 16.11 The bones of the wrist, hand and fingers. Anterior view

 Pot luck

34. Name the bone on which the following landmarks are found:

a. Deltoid tuberosity: _____

b. Acromion process: _____

c. Olecranon process: _____

d. Glenoid cavity: _____ .

 Colouring, matching and labelling

35. On Figure 16.12, colour and match the three fused bones that form the hip bone:

- ○ Ilium
- ○ Ischium
- ○ Pubis

36. Identify the bony landmarks of the hip bone indicated on Figure 16.12.

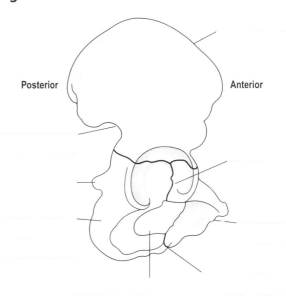

Posterior   Anterior

Figure 16.12 The right hip bone. Lateral view

37. On Figure 16.13, colour and match the following parts of the femur:

○ Neck
○ Head
○ Shaft

38. Label the landmarks of the femur indicated on Figure 16.13.

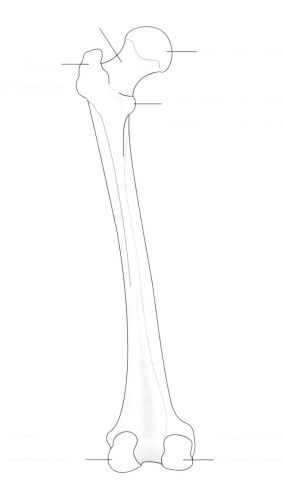

Figure 16.13 The left femur. Posterior view

39. On Figure 16.14, colour and match the following parts of the lower limb:

○ Tibia
○ Fibula
○ Interosseous membrane
○ Area of proximal tibiofibular joint
○ Area of distal tibiofibular joint

40. Label the bony landmarks identified on Figure 16.14.

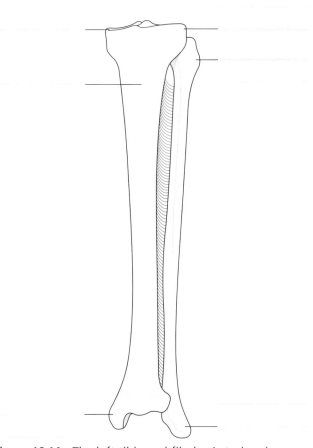

Figure 16.14 The left tibia and fibula. Anterior view

41. What is the function of the interosseous membrane?

_____

_____

42. On Figure 16.15, colour and match the following parts of the foot:

- ○ Tarsal bones
- ○ Metatarsal bones
- ○ Phalanges

43. Label the tarsal bones of the foot shown on Figure 16.15.

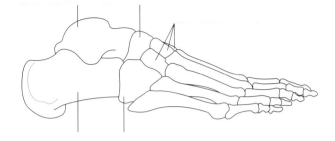

Figure 16.15 The bones of the foot. Lateral view

 Pot luck

44. Define the term 'flat foot' and explain how the structure of the foot changes in this condition.

_____

_____.

JOINTS

Completion

45. Complete the blank columns in Table 16.3 by naming the type of joint listed, using S=synovial, F=fibrous or C=cartilaginous, and identifying the range of movement possible at that joint using I=immovable, Sl=slightly movable and Fr=freely movable.

	Type (S, F or C)	Movement (I, SI, Fr)
Suture		
Tooth in jaw		
Shoulder joint		
Symphysis pubis		
Knee joint		
Interosseous membrane		
Hip joint		
Joint between phalanges		
Intervertebral discs		

Table 16.3 Joints and movements

Matching

46. Match the key choices to define the movements listed below.

Key choices:
Abduction
Adduction
Circumduction
Eversion
Extension
Flexion
Inversion
Pronation
Rotation
Supination

a. Bending, usually forwards: _____

b. Straightening or bending backwards: _____

c. Movement away from the midline of the body: _____

d. Movement towards the midline of the body: _____

e. Movement of a limb or digit so that it forms a cone in space: _____

f. Movement round the long axis of a bone: _____

g. Turning the palm of the hand down: _____

h. Turning the palm of the hand up: _____

i. Turning the sole of the foot inwards: _____

j. Turning the sole of the foot outwards: _____.

 Colouring and labelling

47. Colour and label the following parts of the synovial joint on Figure 16.16:

Bone
Capsular ligament
Synovial membrane
Articular cartilage
Synovial cavity

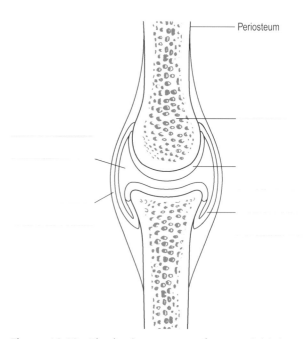

Periosteum

Figure 16.16   The basic structure of a synovial joint

? MCQs

48. Which ONE of the following is NOT a function of synovial fluid? _____.
    a. Nourishment of intracapsular structures
    b. Phagocytic removal of joint debris
    c. Lubrication of joint movement
    d. Reduces friction between the bone ends and joint sleeve.

49. A bursa is a: _____.
    a. Pocket of synovial fluid surrounded by a synovial membrane
    b. Cushioning fat pad
    c. Supporting wedge of cartilage
    d. Localized thickening of synovial membrane.

50. What tissue forms articular cartilage? _____.
   a. A blend of hyaline and fibrocartilage
   b. Fibrocartilage arranged in a double layer for durability
   c. Hyaline cartilage with no blood supply
   d. Elastic cartilage to allow joint movement.

51. Joints are stabilized by (choose all that apply): _____.
   a. Intracapsular ligaments
   b. Smooth muscle and its tendons
   c. Extracapsular ligaments
   d. Wedges of hyaline cartilage (menisci).

   Colouring, matching and labelling

52. Colour and match the following parts of the shoulder joint on Figure 16.17:

○ Capsular ligament
○ Synovial membrane
○ Articular cartilage
○ Synovial cavity

53. Label the parts of the shoulder indicated on Figure 16.17.

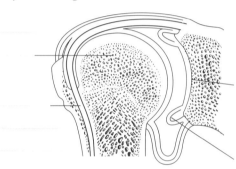

Ⓐ

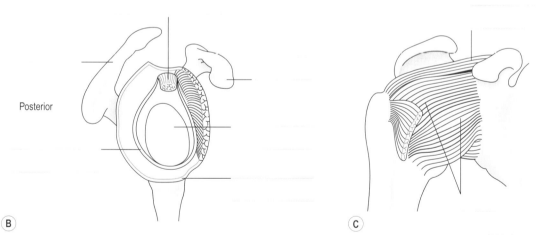

Posterior

Ⓑ   Ⓒ

Figure 16.17 **The right shoulder joint.** A. Section viewed from the front. B. The position of the glenoidal labrum with the humerus removed. C. The supporting ligaments viewed from the front

54. What type of synovial joint is the shoulder joint? _____.

55. The shoulder joint is the most mobile in the body. What relationship exists between the mobility and the stability of a joint?

_____.

Completion

56. Identify the muscles, or combinations of muscles, involved in movements at the shoulder joint and complete Table 16.4.

Movement	Muscle(s) involved
Flexion	
Extension	
Abduction	
Adduction	
Circumduction	
Medial rotation	
Lateral rotation	

Table 16.4 Muscles involved in movement at the shoulder joint

57. What type of synovial joint is found at the elbow? _____.

58. Complete Table 16.5 by inserting the two types of movement possible at the elbow in the left-hand column and the muscles involved in the right-hand column.

Movement	Muscle(s) involved

Table 16.5 Muscles involved in movement of the elbow

59. What type of synovial joint is found between the phalanges? _____.

60. Where else in the body are phalanges found? _____.

Matching

61. Insert the movements from the list of key choices to complete Table 16.6. (Take care, as you will not need all the key choices.)

Key choices:
Abduction
Adduction
Circumduction
Eversion
Extension
Flexion
Inversion
Pronation
Rotation
Supination

Movement of radioulnar joints	Muscle(s) involved
	Pronator teres
	Supinator, biceps
Movement of the wrist	
	Flexor carpi radialis, flexor carpi ulnaris
	Extensor carpi radialis (longis and brevis), extensor carpi ulnaris
	Flexor and extensor carpi radialis
	Flexor and extensor carpi ulnaris

Table 16.6 Muscles involved in movement of the proximal and distal radioulnar joints and wrist

62. What kind of synovial joint is found at the hip?_____.

  Colouring, labelling and matching

63. Colour and match the following parts of the hip joint:

○ Capsular ligament
○ Synovial membrane
○ Articular cartilage
○ Synovial cavity

64. Label the parts of the hip joint indicated in Figure 16.18.

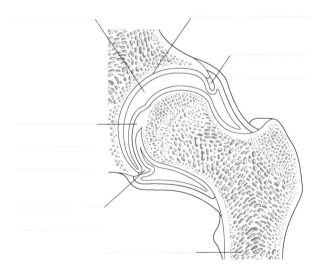

Figure 16.18 Section of the hip joint. Anterior view

Completion

65. Match the groups of muscles in the key choices below with the movements of the hip shown in Table 16.7:

Key choices:
Mainly gluteal muscles and
   adductor group
Gluteus medius and
   minimus, sartorius
Adductor group
Gluteus medius and minimus
Psoas, iliacus, rectus femoris,
   sartorius
Gluteus maximus, hamstrings

Movement	Muscle(s) involved
Flexion	
Extension	
Abduction	
Adduction	
Medial rotation	
Lateral rotation	

Table 16.7 Muscles involved in movement of the hip

 Colouring, matching and labelling

66. Colour and match the following structures of the knee joint on Figures 16.19A and B:

  ○ Articular cartilage
  ○ Capsular ligament
  ○ Cruciate ligament
  ○ Femur
  ○ Fibular
  ○ Patella
  ○ Synovial membrane
  ○ Semilunar cartilages
  ○ Tibia

67. Label the remaining structures shown on Figure 16.19.

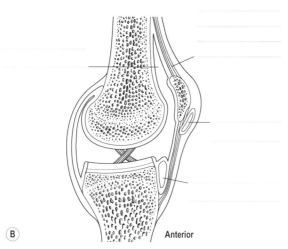

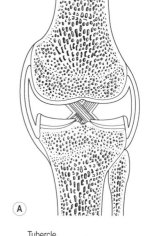

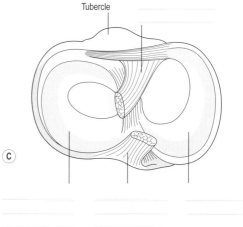

Figure 16.19 The knee joint. **A.** Section viewed from the front. **B.** Section viewed from the side. **C.** The superior surface of the tibia showing the semilunar cartilages and the cruciate ligaments

68. What kind of synovial joint is found at the knee?_____.

Completion

69. Insert the key choices below to complete Table 16.8.

Key choices:
Hamstrings
Quadriceps femoris
Flexion
Gastrocnemius
Extension

Movement	Muscle(s) involved

Table 16.8 Muscles involved in movement of the knee

 Matching

70. List A names some important ligaments. For each, match it with the relevant statement in list B. Be careful, because not all items on list B will be required!

List A

1. Flexor retinaculum _____

2. Cruciate ligament _____

3. Annular ligament _____

4. Plantar ligament _____.

List B

a. Supports the arch of the foot

b. Attaches the head of the femur to the acetabulum

c. Forms the carpal tunnel, through which the tendons of arm muscles run

d. Long ligament running the length of the vertebral column, helping to bind the vertebrae together

e. Holds the odontoid process of the axis in place

f. Forms a ring holding the ulnar and radial heads together

g. Stabilizes the knee joint

h. Holds teeth in place in the jaw.

MUSCLE TISSUE

Pot luck

71. Name the three types of muscle:

- _____

- _____

- _____

Colouring and labelling

72. Colour and label the parts of the skeletal muscle identified on Figure 16.20.

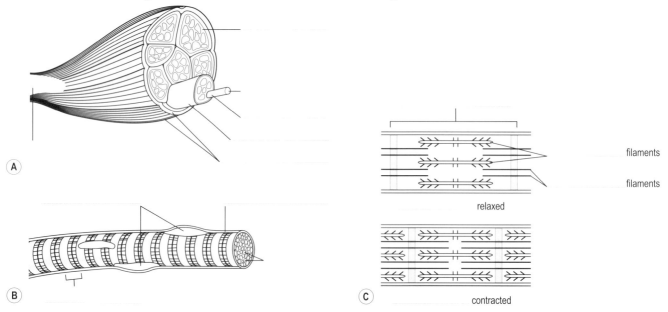

Figure 16.20 Organization within a skeletal muscle

Completion

73. The following paragraph describes the sliding filament theory. Complete it by filling in the blanks.

The functional unit of a skeletal muscle cell is the _____. At each end of this unit are lines called

_____ lines. Within the unit are two types of filament, thick filaments (made of _____), and

thin ones (made of _____). When the muscle cell is relaxed, these two filaments are not connected to

each other. Contraction is initiated when an electrical impulse, called an _____, passes along the cell

membrane (also called the _____) of the muscle cell and penetrates deep into the sarcoplasm via the

network of _____ that run through the cell. This electrical stimulation causes _____ ions to be

released from the _____ within the cell; these ions cause links, called _____ to form between

the thick and thin filaments. The filaments pull on each other, which causes the functional unit to _____

in length, pulling the _____ at either end towards one another. If enough units are stimulated to contract

at the same time, the entire _____ will also _____.

Labelling, matching and colouring

74. Label the parts of the motor unit shown on Figure 16.21.

75. Colour and match the following structures on Figure 16.21:

○ Muscle fibres
○ Nuclei of muscle fibres
○ Motor end plates

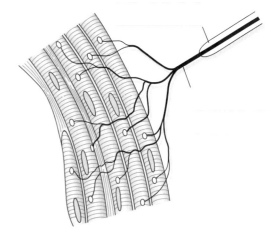

Figure 16.21 The neuromuscular junction

76. Which neurotransmitter is released at the motor end plates? _____.

77. What is a motor unit? _____
_____.

Definitions

Define the terms:

78. Isotonic contraction. _____

_____

79. Isometric contraction. _____

_____

80. The origin of a muscle. _____

_____

81. Hypertrophy of muscle. _____

_____

82. Antagonistic pair. _____

_____

MUSCLES OF THE FACE AND NECK

  Colouring and labelling

83. Colour and label the muscles shown in Figure 16.22.

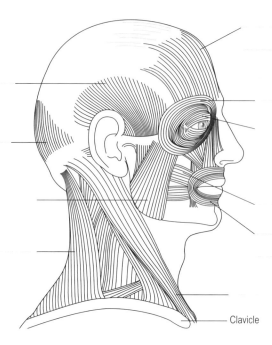

Figure 16.22 The main muscles on the right side of the face, head and neck

Clavicle

 Completion

84. Complete Table 16.9 to outline the functions of the muscles of the face and neck.

Muscle	Paired/unpaired	Function
Occipitofrontalis		
Levator palpebrae superioris		
Orbicularis oculi		
Buccinator		
Orbicularis oris		
Masseter		
Temporalis		
Pterygoid		
Sternocleidomastoid		Contraction of one side: Contraction of both sides:
Trapezius		

Table 16.9 Functions of muscles of the face and neck

MUSCLES OF THE BACK

 Labelling

85. Label the muscles of the back shown on Figure 16.23.

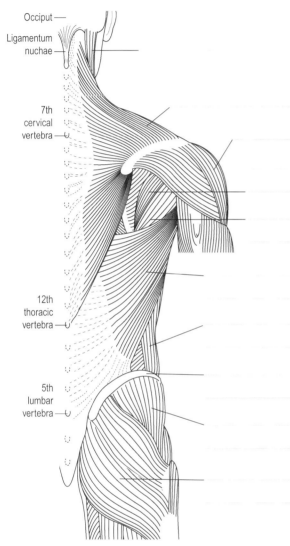

Occiput —

Ligamentum nuchae —

7th cervical vertebra —

12th thoracic vertebra —

5th lumbar vertebra —

Figure 16.23 The main muscles of the back

MUSCLES OF THE ABDOMINAL WALL

   Colouring, matching and labelling

86. Colour and match these muscles on Figures 16.24A and B:

○ Internal oblique

○ External oblique

○ Transversus abdominis

87. Colour and label the other structures shown on Figure 16.24.

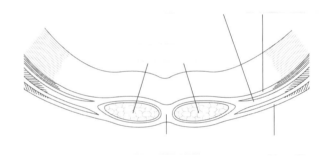

Ⓐ

Anterior aspect

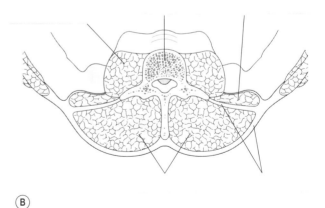

Ⓑ

Figure 16.24 Transverse sections of the muscles and fasciae of the abdominal wall. A. Anterior wall. B. Posterior wall: a lumbar vertebra and its associated muscles

MUSCLES OF THE UPPER LIMBS

  Colouring and labelling

88. Colour and label the muscles of the upper limb on Figure 16.25.

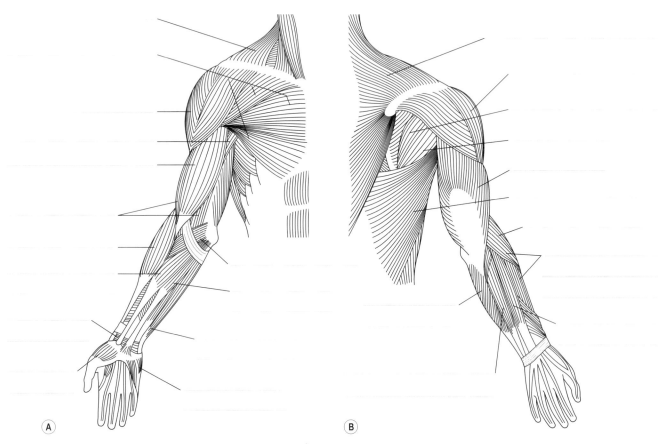

Figure 16.25 The main muscles that move the joints of the upper limb. **A.** Anterior view. **B.** Posterior view

MUSCLES OF THE LOWER LIMBS

  Colouring and labelling

89. Colour and label the muscles of the lower limb shown in Figure 16.26.

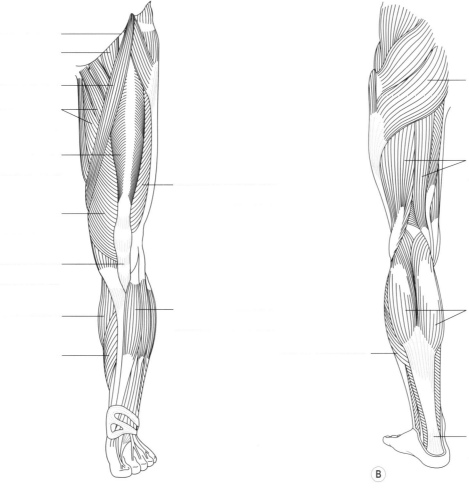

Figure 16.26 The main muscles of the lower limb. A. Anterior view. B. Posterior view

MUSCLES OF THE PELVIC FLOOR

  Colouring and matching

90. Colour and match the following parts of the pelvic floor on Figure 16.27:

- ○ External anal sphincter
- ○ Coccygeus
- ○ Levator ani
- ○ Coccyx
- ○ Vaginal orifice
- ○ Anal orifice
- ○ Urethral orifice

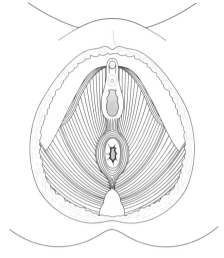

Figure 16.27 The muscles of the female pelvic floor

91. Outline the function of the pelvic floor.

_____

_____

_____.

Introduction to genetics

> Genetics is the study of genes, which direct the function of body cells, and transmit hereditary information from one generation to the next (heredity).

 Pot luck

1. What do the following terms stand for?

   a. DNA _____

   b. RNA _____.

 MCQs

2. A haploid cell has:_____.

   a. 23 chromosomes
   b. No nucleus
   c. 22 pairs of autosomes
   d. Two X-chromosomes.

3. Which of the following describes the structural hierarchy of genetic material, starting with the smallest?_____.

   a. DNA, chromosome, gene, nucleotide
   b. Gene, DNA, nucleotide, chromosome
   c. Nucleotide, DNA, gene, chromosome
   d. Chromosome, nucleotide, gene, DNA.

4. The following statements relate to the sex chromosomes. Which of the four options is NOT true?_____.

   a. The Y chromosome is considerably smaller than the X
   b. The sex chromosomes, also known as the autosomes, are responsible for determining gender
   c. A combination of XX codes for a female child
   d. In the cell karyotype, they are designated chromosome pair 23.

5. Autosomes:_____.

   a. Are chromosomal pairs 1–23
   b. Are present in identical pairs
   c. Contain no genes
   d. Make up the entire karyotype of the cell.

 Completion

6. The paragraph below describes the structure and function of DNA. Complete the paragraph by filling in the blanks.

The nucleus contains the body's _____ material, in the form of DNA, which is built from nucleotides, each

made up of three components: a _____ group, the sugar _____ and one of four _____.

DNA is a double strand of nucleotides that resembles a _____, or twisted ladder. DNA and associated

proteins, also called _____, are coiled together, forming _____. During cell division, the DNA

becomes very tightly coiled and can be seen as _____ under the microscope. There are _____ pairs

of them in most human cells. Each consists of many functional subunits called _____. Any given type of cell uses only part of the whole genetic code, also called the _____, to carry out its specific activities.

Each _____ contains the genetic code, or instructions, for the synthesis of one _____, that could,

for example, be an _____ needed to catalyse a particular chemical _____, a hormone, or it may

form part of the structure of a cell. The coded instructions have to be transferred to the _____ of the cell,

since that is where the organelles that make protein, the _____, are found. DNA itself does not transfer,

but a copy of the genetic code is made in the form of _____, which leaves the _____. When its

instructions have been read and the new protein synthesized, the copy is destroyed.

  ## Matching, colouring and completion

7. Figure 17.1 shows a section of DNA. Colour the sugar and phosphate groups in the backbone different colours.

8. Complete the base sequence on strand 1 in Figure 17.1 by drawing in and colouring the bases to complement strand 2.

9. For the following base sequence in a single strand of DNA, work out the corresponding sequence:

   a. in the other strand of the DNA molecule, and
   b. of a piece of mRNA made from this DNA.

Use this information to complete Table 17.1.

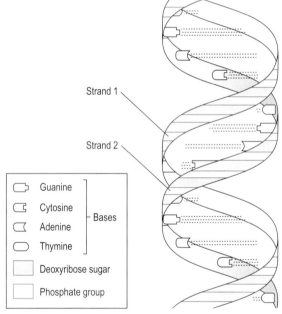

Figure 17.1   Deoxyribonucleic acid (DNA)

DNA Strand 1	C	C	G	T	A	A	C	T	C	A	A	T	G	T
DNA Strand 2														
mRNA														

Table 17.1   The DNA code

Labelling, colouring and matching

10. Figure 17.2 shows the mechanism of protein synthesis from the DNA code. Label the structures indicated.

11. Colour and match the two open arrows to show:

○ transcription
○ translation

12. The list below describes characteristics of the nucleic acids. Decide whether each characteristic applies to DNA, RNA or both by writing DNA, RNA or Both against each item.

a. Contains uracil _____

b. Contains deoxyribose sugar _____

c. Contains phosphate _____

d. Its information is read by translation _____

e. Contains thymine _____

f. Contains ribose sugar _____

g. Is destroyed after its message is read _____

h. Is single stranded _____

i. Contains guanine _____

j. Its information is read by transcription _____.

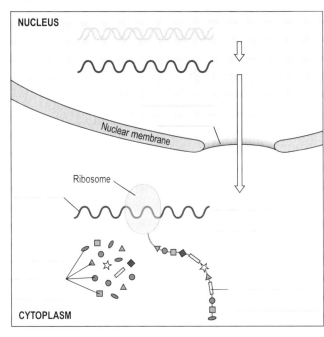

Figure 17.2 The relationship between DNA, RNA and protein synthesis

 Pot luck

13. Of the following eight statements, only four are correct. Identify the incorrect statements and write the corrected version in the spaces provided.

a. Translation takes place in the nucleus

b. The base code in DNA is read in triplets

c. A codon is a piece of DNA carrying information

d. Stop and start codons initiate and terminate protein synthesis

e. All new proteins made by a cell must be used within that cell

f. All body cells contain an identical copy of the genome

g. In each cell, genes whose function is not required are kept switched off

h. Proteins are built on ribosomes in the cytoplasm.

Completion

14. Table 17.2 relates to mitosis and meiosis. Complete it by filling in the appropriate boxes.

	Mitosis	Meiosis
One division or two?		
Daughter cells diploid or haploid?		
Does crossing over take place?		
Are daughter cells identical to parent?		
Two or four cells produced?		
Are daughter cells identical to one another?		
Which process produces gametes?		
Which process replaces damaged cells?		

Table 17.2 Characteristics of mitosis and meiosis

15. The following paragraph relates to autosomal inheritance. Complete it by scoring out the incorrect options in bold.

One chromosome of each pair is inherited from the mother and one from the father, so there are **two/four** copies of each gene in the cell. Two chromosomes of the same pair are called **homologues/homozygotes/autosomes**, and the genes are present in paired sites called **chromatids/traits/alleles**.

When the paired genes are identical, they are called **homozygous/heterozygous**, but if they are different forms they are called **homozygous/heterozygous**. Dominant genes are always **present on the maternal chromosome/ expressed over recessive genes/found in pairs**. Individuals homozygous for a dominant gene **can/cannot** pass the recessive form on to their children, and individuals heterozygous for a gene **can/cannot** pass on either form of the gene to theirs.

Definitions

Define the following terms:

16. Phenotype. _____

_____.

17. Genotype. _____

_____.

Completion

18. Complete the Punnett square (Box 17.1) to illustrate the possible combinations of genes in the children of parents both heterozygous for the ability to roll their tongue. Use T=dominant gene; t=recessive gene.

Box 17.1

	Father's genes	
Mother's genes		

19. Which of the genotypes above will give a tongue-rolling child? _____.

20. Which of the genotypes in Box 17.1 are homozygous? _____.

21. Complete the Punnett square (Box 17.2) to illustrate the possible combinations of genes in the children of a mother homozygous for the recessive gene for blue eyes, and a father homozygous for the dominant gene for brown eyes. Use B=brown eyes, b=blue eyes.

Box 17.2

	Father's genes	
Mother's genes		

22. If the parents have four children, one each of the genotypes above, how many blue-eyed children will they have? _____.

23. Red-green colour blindness is inherited on the X-chromosome (sex-linkage). Complete the Punnett square (Box 17.3), to show the possible combinations of genes in the children of a colour-blind mother and a normally sighted father. Don't forget to include the sex chromosomes (XX and XY); use B=normal gene, b=colour-blind gene as superscripts, e.g. X^B.

Box 17.3

	Father's genes	
Mother's genes		

24. What is the male:female ratio in the children above? _____.

25. What percentage of the boys will be colour blind? _____.

26. What term is used to describe the genetic condition of the girls? _____.

The reproductive system

The reproductive systems in men and women are built differently, although their common function is primarily to ensure production of children and the passing on of the parents' genetic material into another generation. Both systems produce gametes, or sex cells, which fuse to form a potential human being. Females have the additional role of protecting the developing baby within the womb, giving birth and nourishing it in the months after birth. This chapter will test your understanding of the structures and processes involved.

THE FEMALE REPRODUCTIVE SYSTEM

 Labelling

1. Figure 18.1 shows the female external genitalia. Label the structures shown.

2. Indicate the position of the perineal area on Figure 18.1.

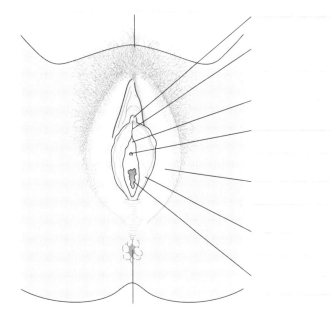

Figure 18.1   Female external genitalia

3. What is the hymen?

_____

4. What is the function of the vestibular glands?

_____

The reproductive system

   Labelling, colouring and matching

5. Figure 18.2 shows the internal female reproductive organs. Label the structures indicated.

6. On Figure 18.2, colour and match the following:

- ⃝ Uterus
- ⃝ Uterine tubes
- ⃝ Ovary
- ⃝ Vagina
- ⃝ Broad ligament

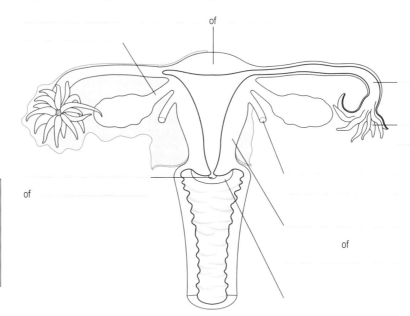

Figure 18.2 The female reproductive organs in the pelvis

7. Figure 18.3 shows the female reproductive organs and some associated structures in the pelvis. Colour and match the following structures:

- ⃝ Bladder
- ⃝ Peritoneum
- ⃝ Pubic bone
- ⃝ Sigmoid colon and rectum
- ⃝ Vertebrae

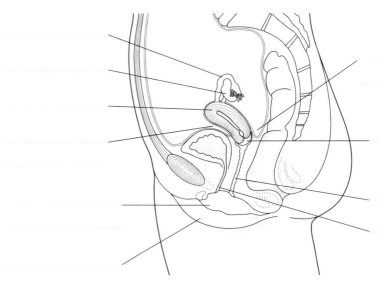

Figure 18.3 Lateral view of the female reproductive organs in the pelvis

8. Label the structures indicated on Figure 18.3.

9. In which of the structures indicated on Figure 18.3 is the pH kept between 4.9 and 3.5, and why?

10. Figure 18.4 shows a section through the uterus. Colour and match the three layers of the uterine wall:

○ Endometrium
○ Myometrium
○ Perimetrium

11. Label the structures shown on Figure 18.4.

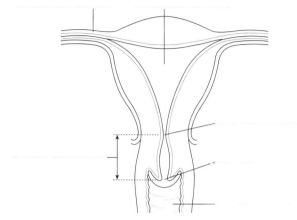

Figure 18.4   A section through the uterus

? MCQs

12. Which layer of the uterine wall is shed during menstruation? _____.

    a. Myometrium   **b.** Functional endometrium   **c.** Perimetrium   **d.** Basal endometrium.

13. Which mechanisms propel the ovum along the uterine tube (choose all that apply): _____.

    a. Peristalsis   **b.** Fimbrial propulsion   **c.** Uterine contractions   **d.** Ciliary movement.

14. Which ligament drapes over the uterine tubes like a blanket, hanging down each side and helping to anchor them in the pelvic cavity? _____.

    a. Round ligament   **b.** Tubular ligament   **c.** Suspensory ligament   **d.** Broad ligament.

15. The ovary secretes: _____.

    a. Oestrogen and luteinizing hormone   **c.** Oestrogen, progesterone and follicle stimulating hormone
    b. Progesterone and oxytocin   **d.** Progesterone and oestrogen.

Labelling, colouring and matching

16. Figure 18.5 shows the main stages of development of a single ovarian follicle. Colour, match and label the following structures as you follow through the process:

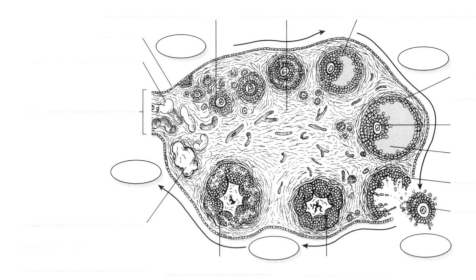

Figure 18.5 Stages of follicular development within the ovary

○ Primordial follicle
○ Follicle approaching maturity
○ Mature ovarian follicle
○ Ovum within follicle
○ Corpus luteum forming
○ Fully formed corpus luteum
○ Fibrous corpus albicans

17. On Figure 18.5, indicate the times within an average ovarian cycle that each of these stages would be reached by completing the time scale in the open ovals around the ovary.

18. Name the process taking place at A. _____.

19. Complete Figure 18.5 by labelling the remaining structures indicated.

Completion

20. The following paragraph describes the control of ovarian function. Score out the incorrect options in bold, leaving the correct one.

Maturation of the follicle is stimulated by **luteinizing hormone/follicle stimulating hormone/ progesterone** released by the anterior pituitary, and oestrogen from the **follicular cells/anterior pituitary/placenta**. Ovulation is triggered by a surge of **luteinizing hormone releasing hormone/luteinizing hormone/oestrogen**, which is secreted by the anterior pituitary. This release takes place a few **minutes/hours/days** before ovulation. After ovulation, the now empty follicle develops into the **primordial follicle/corpus albicans/corpus luteum**, and its main function is to secrete **progesterone/oestrogen/progesterone and oestrogen**, which maintain(s) the uterine lining in case fertilization and implantation occur. If pregnancy does occur, the embedded ovum supports the corpus luteum by producing **human chorionic gonadotrophin/luteinizing hormone/oestrogen**, which keeps it functioning for the next 3 months or so, until the **corpus albicans/placenta/umbilical cord** is developed enough to take on this role. If pregnancy does not occur, the corpus luteum degenerates and forms a scar on the surface of the ovary called the **liquor folliculi/corpus albicans/germinal epithelium**.

 Pot luck

21. List the main changes that take place in the female body during puberty:

- _____
- _____
- _____
- _____
- _____
- _____.

 Completion

22. Figure 18.6 summarizes the main female reproductive hormones and the glands that synthesize them. Complete the figure by filling in the names of the glands (in the boxes) and the hormones a, b, c, d and e.

23. For each of the hormones named in question 21, describe their main functions:

a. _____
_____

b. _____
_____

c. _____
_____

d. _____
_____

e. _____
_____.

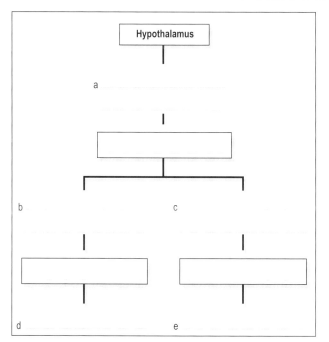

Figure 18.6 Female reproductive hormones and target tissues

Labelling and matching

Figure 18.7 represents various body changes during one reproductive cycle. Each part of the figure summarizes a subcycle involving different tissues and organs. Complete the figure as instructed below.

24. Figure 18.7A summarizes the ovarian cycle. Identify event E.

_____ .

25. Identify the two main hormones and the tissues in the ovary that synthesize them before and after event E by labelling the boxes in Figure 18.7A.

_____ .

26. Figure 18.7B shows the release of follicle stimulating hormone and luteinizing hormone. Which endocrine gland secretes them?

_____ .

27. What is the significance of event E and the peak concentrations of these two hormones occurring simultaneously?

_____ .

28. Figure 18.7C shows the uterine cycle. Label the diagram to show:

> The layers of the endometrium (i, ii and iii)
> The phases of the uterine cycle and the length of time spent in each phase (iv, v and vi)

29. Which hormone maintains the thickened inner layer in the second half of the cycle?

_____ .

○ _____
○ _____

30. Figure 18.7D shows the ovarian hormone cycle. Identify the two hormones that are represented by the curves in the diagram by colouring each line using the key below:

31. This reproductive cycle has not resulted in pregnancy. What is the reason for the falling levels of these two hormones in the second half of the cycle?

_____
_____
_____ .

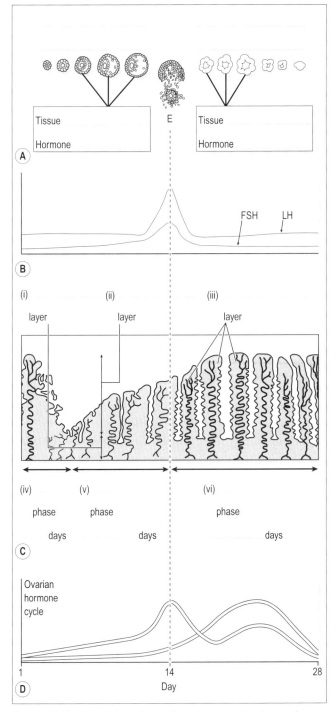

Figure 18.7 Summary of one female reproductive cycle

32. Explain why, should pregnancy occur, the levels of these two hormones would not fall but continue to rise (and remain high during pregnancy).

_____
_____
_____ .

Colouring and labelling

33. Figure 18.8 shows a section through the female breast. Label and colour the structures shown.

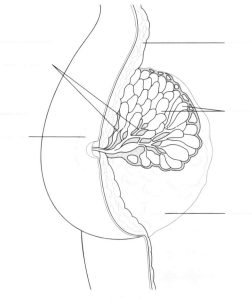

Figure 18.8 Structure of the breast

 Completion

34. Complete Table 18.1 by filling in the appropriate hormone(s) against each statement concerning the breast.

Statement	Hormone(s)
Stimulates body growth and development in puberty	
Initiates release of milk	
Stimulates production of milk	
Stimulates growth and development in pregnancy	

Table 18.1 The effect of hormones on the breast

THE MALE REPRODUCTIVE SYSTEM

  Labelling, colouring and matching

35. Figure 18.9 shows the male reproductive organs and some associated tissues. Using the key below, colour the following structures:

- ○ Peritoneum
- ○ Urinary bladder
- ○ Sigmoid colon and rectum
- ○ Vertebrae

36. Label the other structures shown.

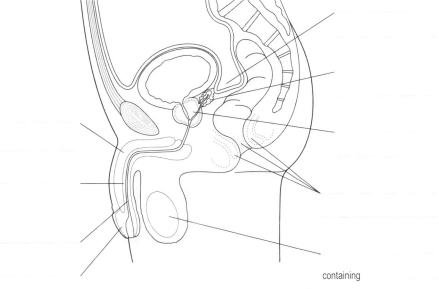

containing

Figure 18.9 The male reproductive organs and their associated structures

221

Labelling, colouring and matching

37. Figure 18.10 shows sections through the testis. Figure 18.10A shows the coverings of the testis, and 18.10B shows the relationship between the testis and the deferent duct (vas deferens). In Figure 18.10A, colour and match the different layers around the testis using the key below:

- ○ Skin
- ○ Smooth muscle
- ○ Tunica vaginalis
- ○ Tunica albuginea
- ○ Connective tissue forming septum

38. In Figure 18.10B, colour, match and label the following sections of tubule:

- ○ Convoluted seminiferous tubules
- ○ Straight seminiferous tubules
- ○ Efferent ductules

39. Colour and label the remaining structures indicated on Figures 18.10A and B.

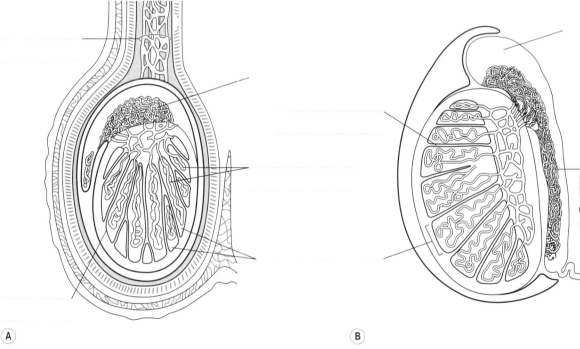

A   **B**

Figure 18.10   The testis. **A.** A section through the testis and its coverings. **B.** A longitudinal section through a testis and its deferent duct

MCQs

40. Testosterone is produced by: _____.

    a. Spermatic cords   **b.** Interstitial cells   **c.** The pituitary gland   **d.** Epididymal cells.

41. The body of the spermatozoon is packed with: _____.

    a. Mitochondria for energy   **c.** Enzymes to penetrate the ovum's coat
    b. DNA for fusion with the ovum   **d.** Motile filaments for propulsion.

42. Which tissue in the penis surrounds the urethra? _____.

    a. The corpus cavernosa   **c.** The corpus spongiosum
    b. The glans penis   **d.** The prepuce.

43. The function of the seminiferous tubules is to: _____.

    a. Produce sperm   **c.** Fill with blood during erection of the penis
    b. Produce seminal fluid   **d.** Store sperm.

44. Which of the following is NOT true of the prostate gland (choose all that apply)? _____.

    a. It stores and matures sperm
    b. It produces a fluid that thickens spermatic fluid in the vagina
    c. It surrounds the urethra and the deferent duct
    d. It lies immediately below the urinary bladder.

  Colouring and matching

45. Figure 18.11 shows a section through the prostate gland and some associated structures. Colour and match the following:

○ Wall of the urinary bladder
○ Ejaculatory duct
○ Urethra
○ Deferent duct
○ Seminal vesicle
○ Prostate gland

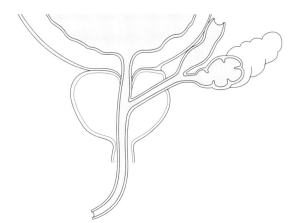

Figure 18.11 Section through the prostate gland and associated structures

Completion

46. Complete Table 18.2.

	% semen volume	Function
Spermatozoa		
Prostate secretions		
Seminal vesicle secretions		

Table 18.2 Nature and function of semen

The body as a whole

ANSWERS

1.

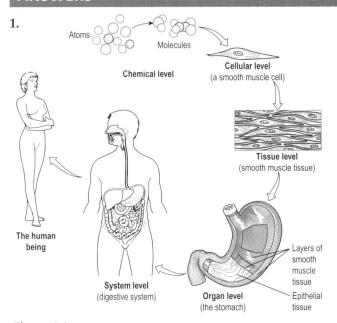

Figure 1.1

3. a.   **4.** c.   **5.** c.   **6.** d.

7. The **external** environment surrounds the body and provides oxygen and the nutrients required by the body cells. The **skin** provides a barrier between the **dry** external environment and the **watery** internal environment. The external environment is the medium in which all body cells exist. Cells are bathed in fluid called **interstitial** fluid, also known as **tissue fluid**. The **cell membrane** provides a potential barrier to substances entering or leaving the cell. **It** prevents **large** molecules moving between the cell and interstitial fluid. **Small** molecules can usually pass through the membrane and therefore the chemical composition of the fluid inside the cell is **different** from that outside. This property is known as **selective permeability**.

2. **Table 1.1** Levels of structural complexity and their characteristics

Level of structural complexity	Characteristics
The human being	Comprises many systems that work interdependently to maintain health
Organ level	Carries out a specific function and is composed of different types of tissue
Cellular level	The smallest independent units of living matter
System level	Consists of one or more organs and contributes to one or more survival needs of the body
Chemical level	Atoms and molecules that form the building blocks of larger substances
Tissue level	A group of cells with similar structures and functions

8.

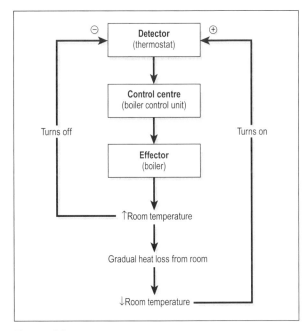

Figure 1.2

9. The composition of the internal environment is maintained within narrow limits, and this fairly constant state is called **homeostasis**. In systems controlled by negative feedback mechanisms, the effector response **reverses** the effect of the original stimulus. When body temperature falls below the preset level, specialized temperature-sensitive nerve endings act as **detectors** and relay this information to cells in the hypothalamus of the brain that form the **control centre**. This results in activation of **effector** responses that raise body temperature. When body temperature returns to the **normal** range again, the temperature-sensitive nerve endings no longer stimulate the cells in the hypothalamus and the heat conserving mechanisms are switched off.

10. Shivering; narrowing of the blood vessels supplying the skin (vasoconstriction).

11. Sweating, widening of the blood vessels supplying the skin (vasodilation).

12. Water and electrolyte concentrations, pH of body fluids, blood glucose levels, blood pressure, blood and tissue oxygen and carbon dioxide levels.

13. It is an amplifier or cascade system where the stimulus progressively increases the response until stimulation ceases.

14. a. Respiratory. b. Digestive. c. Skin (integumentary system).

15. a. Digestive. b. Urinary. c. Respiratory.

16. Non-specific defence mechanisms provide protection against a wide range of invaders, e.g. the skin, mucus from mucous membranes and gastric juice, whereas specific defence mechanisms afford protection against one particular invader (an antigen) and response is through the immune system.

17. a. F, b. T, c. T, d. F, e. T.

18. The childbearing years begin at **puberty** and end at the **menopause**. During this time an **ovum** matures in the ovary about every **28** days. If

fertilization takes place the zygote embeds itself in the **uterus** and grows to maturity during pregnancy, or **gestation**, in about **40** weeks. If fertilization does not occur it is shed with the uterine lining, accompanied by bleeding, called **menstruation**.

19. Carrying to or towards the centre, e.g. central nervous system.

20. Carrying away or away from the centre, e.g. central nervous system.

21. A substance that is recognized as foreign by the immune system, e.g. animal hair, pollen, microorganisms.

22. An abnormally powerful response to an antigen that usually poses no threat to the body.

23.

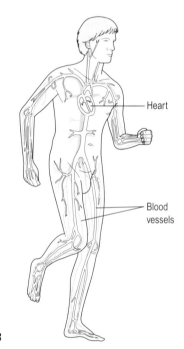

Heart

Blood vessels

Figure 1.3

24. Plasma.

25. 5–6 litres.

26. Arteries.

27. 65–75.

28. and **29.**

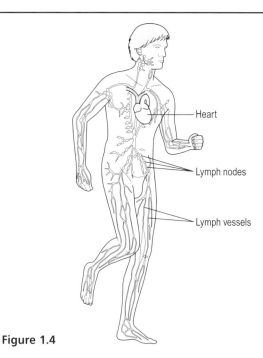

Figure 1.4

30. Filtering of microorganisms and other material from lymph.

31. Lymphocytes.

32. and **33.**

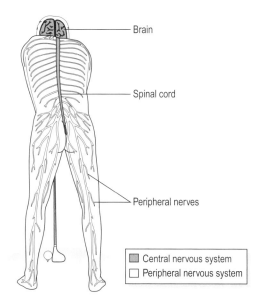

Figure 1.5

34. Reflex action.

35. The endocrine system consists of a number of **glands** in various parts of the body. The glands synthesize and secrete chemical messengers called **hormones** into the **bloodstream**. These chemicals stimulate **target organs/tissues**. Changes in

hormone levels are usually controlled by **negative feedback** mechanisms. The endocrine system, in conjunction with part of the **nervous** system, controls **involuntary** body function. Changes involving the latter system are usually **fast** while those of the endocrine system tend to be **slow** and precise.

36. **Table 1.2** The special senses and their related sensory organs

Special sense	Related sensory organ
Sight	Eye
Hearing	Ear
Balance	Ear
Smell	Nose
Taste	Tongue

37.

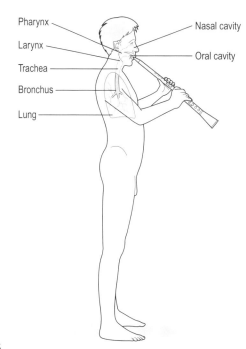

Figure 1.6

38. and **39.** Accessory organs labelled in bold.

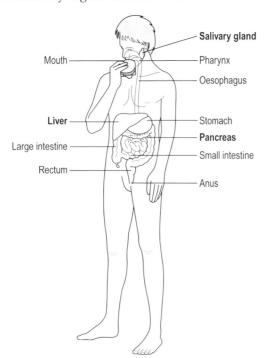

Salivary gland

Mouth — Pharynx

Oesophagus

Liver — Stomach

Large intestine — **Pancreas**

Rectum — Small intestine

Anus

Figure 1.7

40. Enzymes.

41.

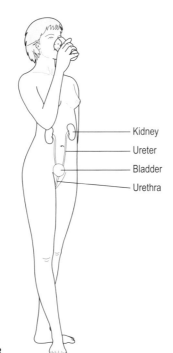

Kidney

Ureter

Bladder

Urethra

Figure 1.8

42. Bladder.

43. Hormones.

44.

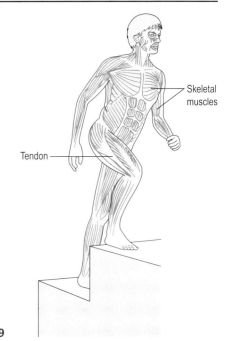

Skeletal muscles

Tendon

Figure 1.9

45.

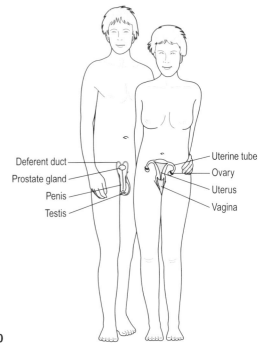

Deferent duct — Uterine tube

Prostate gland — Ovary

Penis — Uterus

Testis — Vagina

Figure 1.10

46. Building up or synthesis of large and complex chemical substances.

47. Breaking down of large chemical substances.

48. Elimination of urine, voiding.

49. Elimination of faeces.

50. a. Symptom. b. Congenital. c. Acquired. d. Chronic. e. Syndrome. f. Acute. g. Sign.

51. The cause of disease.

52. The nature of a disease and its effect on body functioning.

53. The likely outcome of a disease.

54. A disease or condition of which the cause is unknown.

Electrolytes and body fluids

ANSWERS

1. The smallest particle of an element that exists in a stable form.

2. A substance containing one or more elements.

3. A chemical containing only one type of atom.

4. and 5.

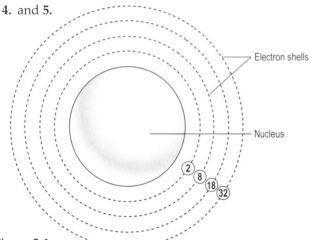

Figure 2.1

6. Electrons: d., e., i.
   Protons: a., b., c., f., g.
   Neutrons: b., c., g., h.

7.

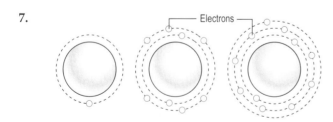

	Hydrogen	Oxygen	Sodium
Atomic number	1	8	11
Atomic weight	1	16	23

Figure 2.2

8. c.   9. b.   10. d.   11. c.

12. The commonest type of atomic bond is called a **covalent** bond. This is a **stable** bond, joining atoms **firmly** together, and the molecules formed are **electrically neutral**. This bond is formed when **electrons** are **shared between atoms**. An example of a molecule with such a bond is **water**.

    The next most common form of atomic bond is the **ionic** bond. This is **less** stable than the bond identified above. It is formed when **electrons** are **donated from one atom to another**. When molecules containing this type of bond are dissolved in water, the bonds break to release **ions**. Such a substance is called an **electrolyte**.

13. Electrolytes conduct electricity, exert osmotic pressure and function in acid–base balance.

14. c.   15. b.   16. d.   17. d.

18. Lungs and kidneys.

19. Abnormally high (alkaline) pH of body fluids.

20. CO_2 (carbon dioxide) + H_2O (water) $\leftrightarrow$ H_2CO_3 (carbonic acid) $\leftrightarrow$ H^+ (hydrogen ion) + HCO_3^- (bicarbonate ion).

21. Acidosis: b., c., e.
    Alkalosis: a., d., f.

22. Table 2.1 Characteristics of some important biological molecules

	Carbohydrates	Proteins	Nucleotides	Lipids
Building blocks are amino acids		✓		
Contain carbon	✓	✓	✓	✓
Molecules joined with glycosidic linkages	✓			
Used to build genetic material	✓		✓	
Building blocks are monosaccharides	✓			
Contain glycerol				✓
Contain hydrogen	✓	✓	✓	✓
Molecules joined together with peptide bonds		✓		
Strongly hydrophobic				✓
Built from sugar unit, phosphate group and base			✓	
Enzymes are made from these		✓		
Contain oxygen	✓	✓	✓	✓

23.

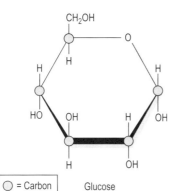

$\bigcirc$ = Carbon   Glucose

Figure 2.3

24. Source of energy for immediate use by cells; energy storage for body; used to build nucleic acids; act as cell surface receptors.

25.

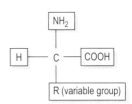

Figure 2.4

26. Insulin; haemoglobin; antibodies; enzymes; collagen.

27.

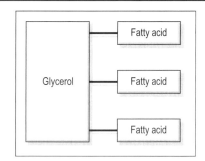

Figure 2.5

28. Lipids are a group of substances that are all strongly hydrophobic, meaning they dissolve **poorly, if at all** in water. Although the different types of lipid are chemically diverse, they all contain **carbon, hydrogen and oxygen** atoms. There are different types of lipid. One type is the fat, stored in the body's adipose tissues as a source of energy. Fat is a **less** efficient source of energy than carbohydrate, as it releases **less** energy when broken down. In addition, fat stored under the skin insulates the body and protects underlying structures. A fat molecule consists of **three** fatty acid molecules linked to a molecule of **glycerol**. Other types of lipid include certain vitamins, e.g. vitamin **A**, **D**, **E** or **K**, an example of a fat-soluble vitamin. Some hormones are lipids, including the steroid hormones. Other lipids include the phospholipids, which form an integral part of the cell's **membrane**.

29.

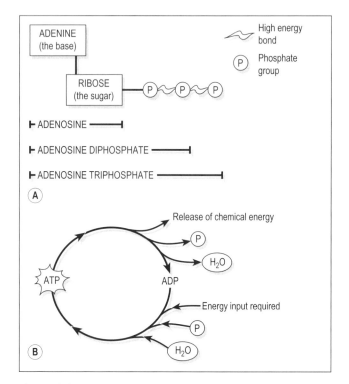

Figure 2.6

30. a. Enzymes are proteins used in the body to <u>speed up</u> chemical reactions in body cells.
    b. T.
    c. Enzymes can catalyse the production of larger molecules from smaller ones, and this is called synthesis or <u>anabolism</u>.
    d. Enzymes are <u>specific</u>. Each enzyme therefore is capable of catalysing <u>one</u> type of reaction.

31. b.

32. a.

33. d.

34. a., b., d.

35. 60%.

36. Cytoplasm, potassium, ATP.

The cells, tissues and organization of the body

CHAPTER
3

ANSWERS

1. and **2.**

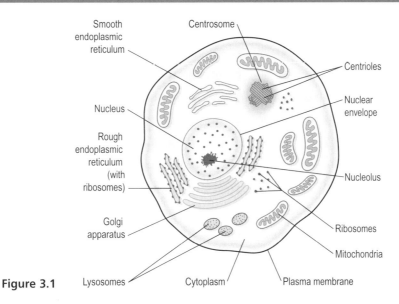

Figure 3.1

3. Table 3.1 Intracellular organelles and their functions

Organelle	Function
Nucleus	The largest organelle, directs the activities of the cell
Mitochondria	Sausage-shaped structures often described as the powerhouse of the cell. Sites of aerobic respiration
Ribosomes	Tiny granules consisting of RNA and protein that synthesize proteins for use within cells
Rough endoplasmic reticulum (ER)	Proteins exported from cells are manufactured here
Smooth endoplasmic reticulum	Lipids and steroid hormones are synthesized here
Golgi apparatus	Stacks of closely flattened membranous sacs that form membrane-bound granules called secretory vesicles
Lysosomes	Secretory vesicles that contain enzymes for the breakdown of large cellular wastes, e.g. fragments of old organelles
Microfilaments	The tiny strands of protein that provide the structural support and shape of a cell
Microtubules	Contractile proteins involved in movement of cells and of organelles within cells

4. The plasma membrane consists of two layers of phospholipids with some **protein** molecules embedded in them. The phospholipid molecules have a head which is electrically charged and hydrophilic (meaning water **loving**) and a tail that has no charge and is hydrophobic. The **lipid** cholesterol is also present. The phospholipid bilayer is arranged like a sandwich with the hydrophilic heads on the **outside** and the hydrophobic tails on the **inside**. These differences influence the passage of substances across the membrane. In motile cells, **extensions** project from the plasma membrane and include long **flagella** which permit movement of the cell.

5., 6. and 7. See Figure 3.2.

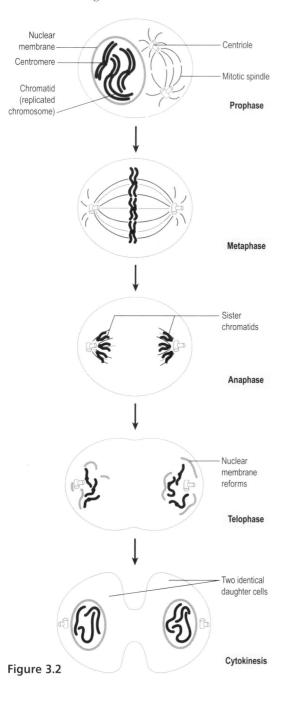

Figure 3.2

8. Most body cells have **46** chromosomes and divide by **mitosis**. The daughter cells of mitosis are genetically **identical.** Formation of gametes takes place by **meiosis** and the daughter cells are genetically **different**. The period between two cell divisions is known as the **cell** cycle, which has two stages, the M phase and interphase. **Interphase** is the longer stage. Interphase has **three** separate stages. The stage where there is most cell growth is the **first gap phase.** The stage at which chromosomes replicate is the **S phase**. Mitosis has **four** identifiable stages.

9. Transport up a concentration gradient that requires chemical energy (ATP).

10. Transport down a concentration gradient without the use of chemical energy (ATP).

11. a., d.   12. b.   13. c.   14. a.   15. a., c., d.   16. b.

17. Transfer of large particles across the plasma membrane into the cell occurs by **phagocytosis** and **pinocytosis**. The particles are engulfed by extensions of the **plasma membrane** that enclose them forming a membrane-bound **vacuole**. Then lysosomes adhere to the cell membrane releasing **enzymes** that **digest** the contents. Extrusion of waste materials by the reverse process is called **exocytosis.**

18. and **19.** All cells are shaded, nuclei within them darkly shaded.

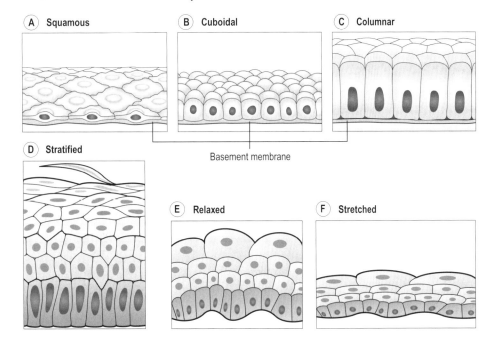

Figure 3.3 (A–F)

20. Urinary bladder.

21. Allows stretching of the bladder as it fills with urine.

22., 23. and **24.**

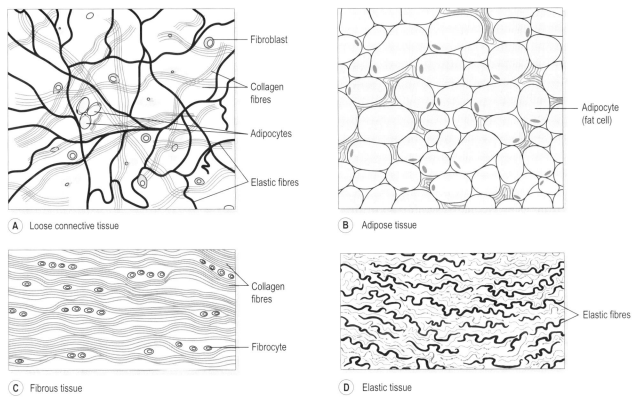

Figure 3.4 (A–D)

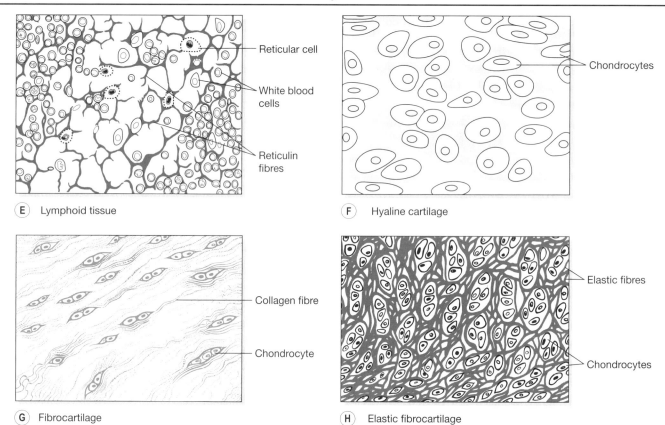

E) Lymphoid tissue

Reticular cell

White blood cells

Reticulin fibres

F) Hyaline cartilage

Chondrocytes

G) Fibrocartilage

Collagen fibre

Chondrocyte

H) Elastic fibrocartilage

Elastic fibres

Chondrocytes

Figure 3.4 (E–H)

25., 26. and **27.** See Figure 3.5.

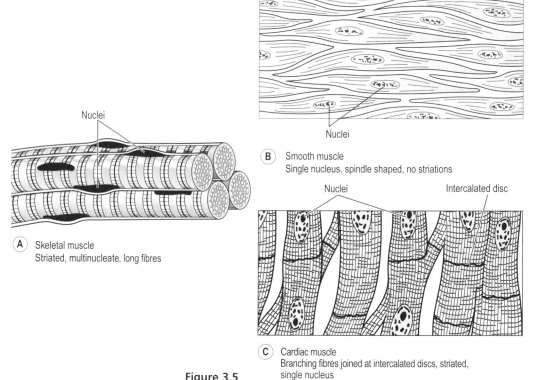

Nuclei

B) Smooth muscle
Single nucleus, spindle shaped, no striations

Nuclei

A) Skeletal muscle
Striated, multinucleate, long fibres

Nuclei

Intercalated disc

C) Cardiac muscle
Branching fibres joined at intercalated discs, striated, single nucleus

Figure 3.5

28. c. **29.** d. **30.** a., b., c., d. **31.** a. **32.** b.

33. c. **34.** a., b., c. **35.** b., c., d.

36. Muscle cells are also called **fibres**. Muscle tissue has the property of **contractility** that brings about movement, both within the body and of the body itself. This requires a blood supply to provide **oxygen, calcium** and **nutrients,** and to remove **wastes**. The chemical energy needed is derived from **ATP.**

Skeletal muscle is also known as **voluntary** muscle because **contraction** is under conscious control. When examined under the microscope, the cells are roughly **cylindrical** in shape and may be as long as **35 cm**. The cells show a pattern of clearly visible stripes, also known as **striations**. Skeletal muscle is stimulated by **motor nerve** impulses that originate in the brain or spinal cord and end at the **neuromuscular junction.**

Smooth muscle has the intrinsic ability to **contract** and **relax**, but it can also be stimulated by **autonomic nerve** impulses, some **hormones** and **local metabolites**.

Cardiac muscle is found only in the wall of the **heart**, which has its own **pacemaker** system, meaning that this tissue contracts in a co-ordinated manner without external stimulation. **Autonomic nerve** impulses and some **hormones** influence activity of this type of muscle.

37. Mucous membrane is sometimes referred to as the **mucosa.** It forms the moist lining of body tracts, e.g. the **alimentary, respiratory** and **genitourinary** tracts. The membrane consists of **epithelial** cells, some of which produce a sticky secretion called **mucus.** This substance protects the lining from **injury.** In the alimentary tract it **lubricates** the contents and in the respiratory system it traps **inhaled particles.**

A serous membrane may also be known as the **serosa.** It consists of a double layer of **loose areolar** connective tissue lined by **simple squamous** epithelium. The layer lining the body cavity is the **parietal** layer and that surrounding organs within a cavity, the **visceral** layer. There are three sites where serous membranes are found:

a. the **pleura** lining the thoracic cavity and surrounding the lungs

b. the **pericardium** lining the pericardial cavity and surrounding the heart

c. the **peritoneum** lining the abdominal cavity and surrounding the abdominal organs.

Synovial membrane lines the cavities of **moveable (synovial) joints**. It consists of **areolar connective** tissue containing **elastic** fibres. This membrane secretes a clear, sticky, oily substance known as **synovial fluid**. It provides **lubrication** and **nourishment**, and prevents **friction** between structures in **synovial** joints.

38. and **39.**

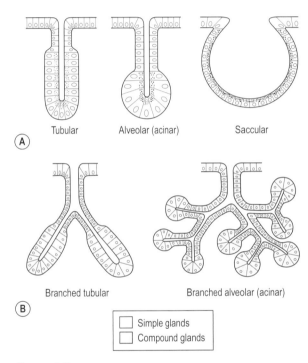

Figure 3.6

40. This is the position assumed in all anatomical descriptions to ensure accuracy and consistency. The body is in the **upright** position with the head facing **forwards**, the arms facing **forwards** with the palms of the hands facing **forwards** and the feet **together.**

41. and **42.**

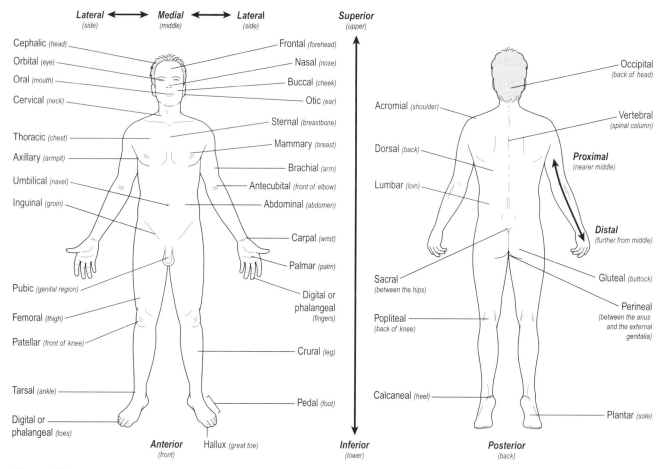

Figure 3.7

43. and **44.**

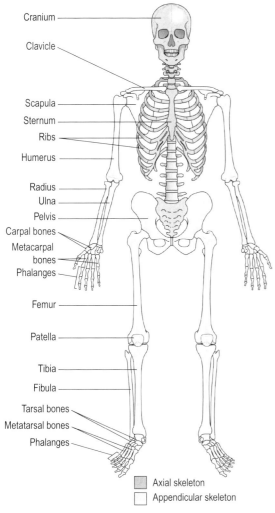

Figure 3.8

49.

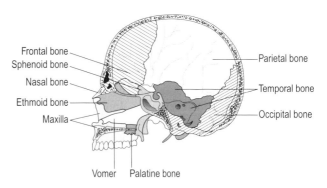

Figure 3.9

50.

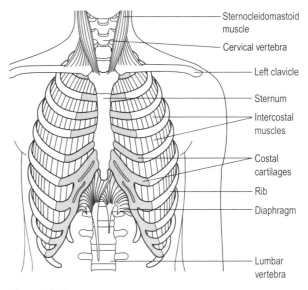

Figure 3.10

45. a. **46.** d. **47.** c. **48.** b.

51. The space between the lungs including the structures situated there, e.g. the heart, oesophagus and blood vessels.

52. The liver.

53. Tranverse colon.

54. and **55.**

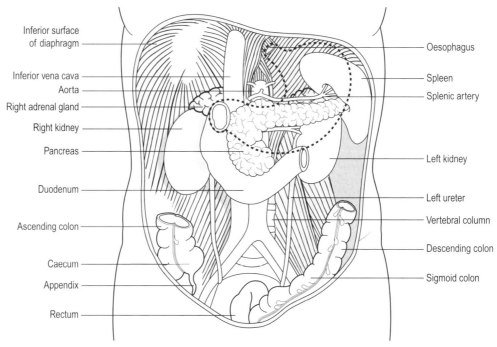

Figure 3.11

56.

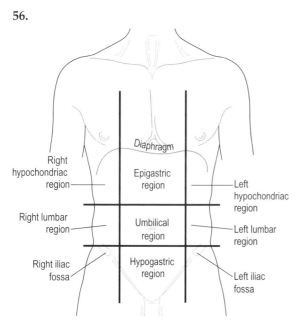

Figure 3.12

57. a. 5; **b.** 3; **c.** 1, 2, 3; **d.** 1, 2, 3, 4; **e.** 5; **f.** 3; **g.** 1; **h.** 1; **i.** 5.

58. a. Abdominal. **b.** Thoracic. **c.** Pelvic. **d.** Abdominal. **e.** Cranial. **f.** Pelvic. **g.** Thoracic. **h.** Abdominal.

59. An agent that can cause malignant changes in cells.

60. A mass of tissue that has escaped the body's normal growth control mechanisms and usually grows faster than normal.

61. a. Fibroblasts. **b.** Adipocytes. **c.** Macrophages. **d.** Plasma cells. **e.** Mast cells.

62. a. The humerus is **lateral** to the heart. **b.** The vertebrae are **posterior** to the kidneys. **c.** The phalanges are **distal** to the ulna. **d.** The skull is **superior** to the vertebral column. **e.** The greater omentum is **anterior** to the small intestine. **f.** The appendix is **inferior** to the stomach. **g.** The patella is **proximal** to the tarsal bones. **h.** The scapulae are **lateral** to the sternum. **i.** The kidneys are **inferior** to the adrenal glands.

63. b. **64.** b. **65.** b. **66.** d.

The blood

ANSWERS

1. and **2.**

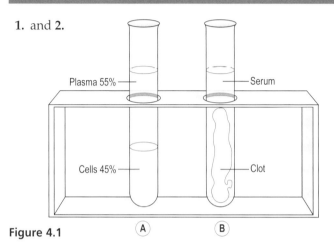

Plasma 55%

Serum

Cells 45%

Clot

Figure 4.1

Ⓐ   Ⓑ

3. Clotting proteins.

4. a. Antibodies; b. Calcium; c. Hormones;
d. Fibrinogen; e. Water; f. Glucose; g. Urea;
h. Bicarbonate; i. Iron; j. Albumins.

5. a. 2, 3, 4; b. 4; c. 3; d. 3, 4; e. 1; f. 4; g. 2; h. 3.

6. and **7.**

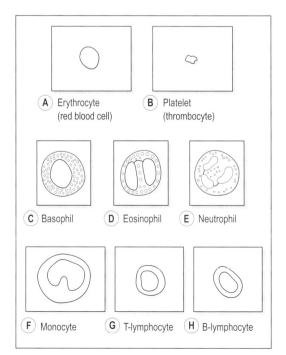

Ⓐ Erythrocyte
(red blood cell)

Ⓑ Platelet
(thrombocyte)

Ⓒ Basophil

Ⓓ Eosinophil

Ⓔ Neutrophil

Ⓕ Monocyte

Ⓖ T-lymphocyte

Ⓗ B-lymphocyte

Figure 4.2

8. Haemopoiesis.

9. To make more space for haemoglobin.

10. Enzymes and toxic chemicals (neutrophils and
eosinophils); histamine and heparin (basophils).

11. a: 11, 12, 13, 17, 18, 19, 20; b: 3, 8, 12, 17, 18; c: 1, 4,
17, 18; d: 4, 7, 14, 17, 18; e: 4, 6, 17, 18; f: 5, 9, 16,
17, 18; g: 5, 9, 10, 15, 17, 18; h: 2, 5, 9, 10, 15, 17, 18.

12. b.   **13.** d.   **14.** a.   **15.** c.   **16.** d.   **17.** b.

18.

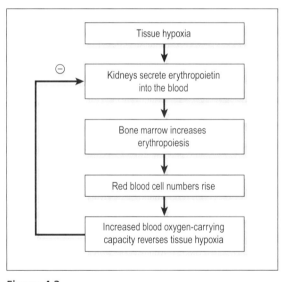

Tissue hypoxia

⊖

Kidneys secrete erythropoietin
into the blood

Bone marrow increases
erythropoiesis

Red blood cell numbers rise

Increased blood oxygen-carrying
capacity reverses tissue hypoxia

Figure 4.3

19. The lifespan of red blood cells is usually
about **120** days. Their breakdown, also called
haemolysis, is carried out by phagocytic
reticuloendothelial cells found mainly in the
liver, **spleen** and **bone marrow**. Their breakdown
releases the mineral **iron**, which is kept by the
body and stored in the **liver**. It is used to form new
haemoglobin. The protein released is converted to
the intermediate **biliverdin**, and then to the yellow
pigment **bilirubin**, before being bound to plasma
protein and transported to the **liver**, where it is
excreted in the **bile**.

20. See Table 4.1 The ABO system of blood grouping.

Blood group	Type of antigen present on red cell surface	Type of antibody present in plasma	Can safely donate to:	Can safely receive from:
A	A	Anti-B	A, AB	A, O
B	B	Anti-A	B, AB	B, O
AB	A, B	Neither	AB	AB, A, B, O
O	Neither	A, B	O, A, B, AB	O

Table 4.1 The ABO system of blood grouping

21. O. **22.** AB.

23. Harold – group A; Olivia – group AB; Alex – group B; Amanda – group O.

24. Harold is blood group A, with A antigens on his red cells and anti-B antibodies in his plasma. This means he cannot receive from any individual with B antigens on their red cells, i.e. individuals with blood groups AB (Olivia) or B (Alex), because he will make antibodies to their red cells. Only Amanda (group O) or another type A individual could be considered for donation.

25. See Table 4.2 Characteristics of white blood cells.

	Neutrophils	Eosinophils	Basophils	Monocytes	Lymphocytes
Phagocyte	✓	✓			
Involved in allergy		✓	✓		
Converted to macrophages				✓	
Release histamine			✓		
Many in lymph nodes					✓
Kupffer cells				✓	
Increased numbers in infections	✓	✓	✓	✓	✓
Kill parasites		✓			
Part of the reticuloendothelial system				✓	

Table 4.2 Characteristics of white blood cells

26.

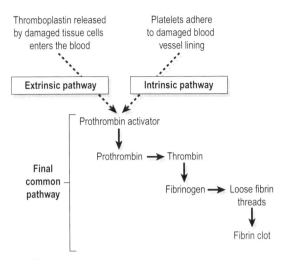

Figure 4.5

27. a. T.
b. The extrinsic and intrinsic pathways both activate the first step in the final common pathway, which is the conversion of <u>prothrombin to thrombin</u>.
c. Clotting factors circulate in the bloodstream in an <u>inactive form, to prevent inappropriate</u> activation of clotting.
d. Prothrombin is clotting factor <u>II</u>.
e. <u>Plasmin</u> is the principal enzyme involved in the breakdown of clots once bleeding has been stopped.
f. T.
g. The platelet plug formed rapidly following blood vessel damage is a <u>temporary structure containing no fibrin</u>.
h. Blood clotting is an example of a <u>positive</u> feedback mechanism.

ANSWERS

1. The heart pumps blood into two separate circulatory systems, the **pulmonary** circulation and the **systemic** circulation. The **right** side of the heart pumps blood to the lungs, whereas the **left** side of the heart supplies the rest of the body. The **capillaries** are the sites of exchange of nutrients, gases and wastes. Tissue wastes, including carbon dioxide, pass into the **bloodstream** and the tissues are supplied with **oxygen** and **nutrients**.

2.

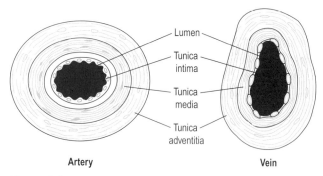

Lumen
Tunica intima
Tunica media
Tunica adventitia

Artery   Vein

Figure 5.1

3. **Table 5.1** Layers of vessel wall

Descriptive phrase	Layer (tunica) of vessel wall
Squamous epithelium	Inner layer (tunica intima)
Contains mainly fibrous tissue	Outer layer (tunica adventitia)
Endothelial layer	Inner layer (tunica intima)
Consists partly of muscle tissue	Middle layer (tunica media)
The vessel's elastic tissue is here	Middle layer (tunica media)
Outer layer	Tunica adventitia

4. c.   5. d.   6. b.   7. c.   8. a.

9. The tiniest arterioles split up into a large number of tinier vessels called **capillaries**. Across the walls of these vessels, the tissues obtain **oxygen** and **nutrients**, and get rid of their **wastes**. The walls of these vessels are therefore thin, being only **one cell** thick. Substances such as **water** and **glucose** can pass across them, whereas larger constituents of blood such as **blood cells** and **plasma proteins** are retained within the vessel. This vast network of microscopic vessels has a diameter of only about **7 μm**, and links the arterioles to the **venules**. In some parts of the body, such as the liver, the vessels in the tissues are wider than this, and are called **sinusoids**. Blood flow here is **slower** than in other tissues because of the bigger lumen.

10. Vasoconstriction.

11. Since sympathetic stimulation generally constricts blood vessels, it is important to protect blood flow to vital organs, e.g. brain and kidneys, even when sympathetic activity is high, such as in the flight-or-fight response.

12. Length of vessel, thickness (viscosity) of the blood and diameter of vessel.

13. Diameter of the vessel.

14. Autoregulation is the **ability of a tissue to control its own blood flow**. It is regulated **locally**. Two examples of autoregulation are **the increase of blood flow through the gastrointestinal tract after eating** and **the reduction in blood flow through the skin when cold**. Another example is the **increase** in blood flow through an active tissue such as exercising skeletal muscle. Blood vessels supplying an active muscle are **dilated** by **hypoxia** in the tissue, and this leads to an **increased** blood supply to match tissue needs.

15. Table 5.2 Characteristics of osmosis, diffusion and active transport

	Osmosis	Diffusion	Active transport
Movement only down a concentration gradient	✓	✓	
Movement of water molecules	✓		
Movement across a semipermeable membrane	✓	✓	✓
Movement requires energy			✓
Movement up a concentration gradient possible			✓
Movement does not require energy	✓	✓	
Movement of oxygen		✓	
Movement of carbon dioxide		✓	

16. a. Osmotic pressure; b. blood pressure; c. hydrostatic (blood) pressure; d. hydrostatic (blood) pressure; e. osmotic pressure.

17.

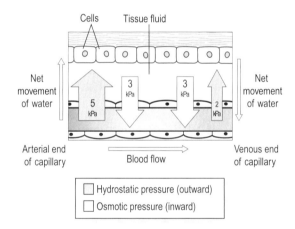

Figure 5.2

18. Because there are a vast number of capillaries, the pressure of the blood drops as it flows through them from the arterioles.

19. It is drained away by the lymphatic system.

20. and 21.

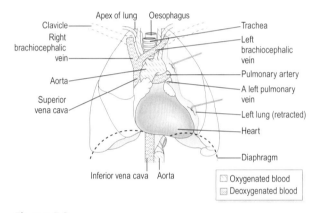

Figure 5.3

22.

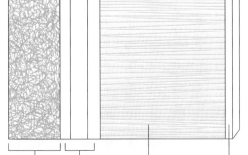

Figure 5.4

23. Table 5.3 Features of heart wall structure

Statement	Layer
Lines the heart chambers	Endocardium
Consists of cardiac muscle	Myocardium
One cell thick	Endocardium
Lines the fibrous pericardium	Parietal pericardium
Outer supportive layer	Fibrous pericardium
Secretes pericardial fluid	Serous pericardium
Contains pericardial fluid	Pericardial space
Lies between the visceral and serous pericardial layers	Pericardial space
Lies between the endocardium and the visceral pericardium	Myocardium
The thickest layer of the heart wall	Myocardium
Inelastic layer	Fibrous pericardium
Covers the heart valves	Endocardium
Covers the myocardium	Visceral pericardium

24. b.   **25.** a.   **26.** b.   **27.** a.

28. Lungs, which are covered with the pleural membrane, and the peritoneal cavity, lined with the peritoneum.

29. and **30.**

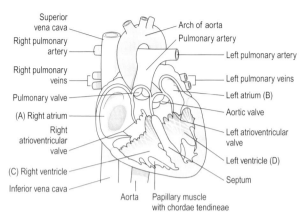

Figure 5.5

31. The chordae tendineae are tendinous cords that fasten the valves to the papillary muscles, and prevent the valves from being pushed into the atria when the ventricles are contracting.

32. The valves prevent backflow of blood in the heart.

33. and **34.**

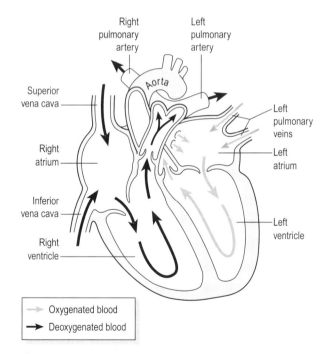

Figure 5.6

35. Aorta, systemic arterial network, capillaries of body tissues, systemic venous network, venae cavae, right atrium, right atrioventricular (tricuspid) valve, right ventricle, pulmonary valve, pulmonary arteries, lungs, pulmonary veins, left atrium, left atrioventricular (mitral) valve, left ventricle, aortic valve, aorta.

36. Because the left ventricle has to pump blood around the systemic circulation, whereas blood from the right is only going as far as the lungs.

37. and 38.

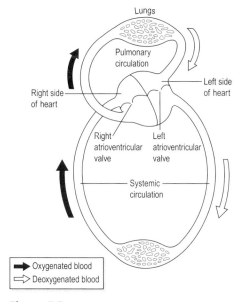

Figure 5.7

39.

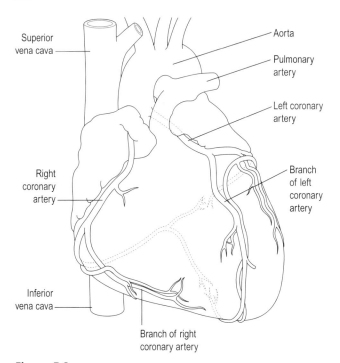

Figure 5.8

40. Supply of blood to the heart wall.

41. The aorta (C).

42. Myocardial ischaemia (angina) or myocardial infarction (heart attack).

43.

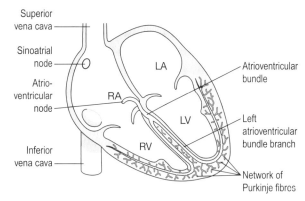

Figure 5.9

44. c. 45. a. 46. c. 47. b.

48.

Diastole

We will begin this description with the heart in diastole, when the whole heart is **resting**. During this time, in the upper part of the heart, the atria are **relaxed** and blood is flowing **into the atria**. Not only the upper chambers are filling but also the lower ones; because the **atrioventricular** valves are open, we see that the **ventricles** are filling as well. Although blood is travelling into the lower chambers, at this stage the electrical activity has not reached them yet so the ventricles are **relaxed**. Remember, during this period, the heart muscle is not contracting; both the **atria** and the **ventricles** are relaxed.

Atrial systole

The next stage represents atrial systole, or contraction. This is initiated when the **sinoatrial node** fires; its electrical discharge leads to the spread of **electrical impulses** through the atria. Because of the electrical excitation of the muscle, the atria **contract** and this leads to pumping of blood from the **atria** into the **ventricles**. It is important therefore that the **atrioventricular** valves are open to permit blood to flow through. The ventricles fill up; because the **aortic** and **pulmonary** valves are closed, blood cannot yet pass from the heart into the great vessels leaving it.

Ventricular systole

The third stage is ventricular systole. The impulse from the sinoatrial node has passed through the atrioventricular node; inspection of the atria shows that they are **relaxed** after their period of activity; this allows them to rest. However, as far as the lower chambers are concerned, because **electrical impulses** are spreading through the **ventricles**, we see that the ventricles **contract**. So that blood cannot flow in a backwards manner into the atria, the **atrioventricular** valves are closed. However, for the ventricles to be able to push blood out of the heart, the **aortic** and **pulmonary** valves **open**. Because of the force generated by the contracting ventricular muscle, blood is pumped from the ventricles into the **pulmonary arteries** and the **aorta**.

The cycle is now complete; the heart will enter another period of diastole, allowing the entire organ to rest briefly before the next period of contraction.

49. Statements C and D are true.

Initial statement		Reason
A. When the ventricles contract, the atrioventricular valves close,	because	rising pressure in the ventricles forces them shut. There is no muscle in the valves.
B. During ventricular contraction, the aortic valve is open,	because	rising pressure in the ventricles forces it open to allow blood to leave.
E. There is a very brief delay between atrial contraction and ventricular contraction,	because	transmission of the electrical signal slows down very slightly as it passes through the atrioventricular (AV) node.
F. Blood does not flow backwards in the heart, i.e. from ventricles to atria,	because	the heart valves prevent it.

50.

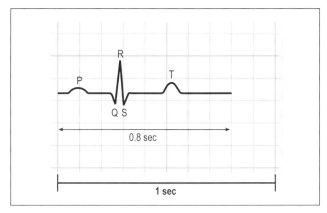

Figure 5.10

51. a. QRS complex; b. P wave; c. T wave; d. QRS complex; e. P wave; f. QRS complex; g. QRS complex; h. QRS complex.

52. 5.6 L.   **53.** 80 ml.   **54.** 100 beats per minute.

55. 120 beats per minute.

56. Heart rate of 60/1.2=50, so this is bradycardia.

57. a., b., g.   **58.** c., d., f., g., h.

59. c.   **60.** b.   **61.** a.   **62.** c.

63. The baroreceptor reflex is important in the **moment-to-moment** control of blood pressure. It is controlled by the cardiovascular centre found in the **medulla oblongata**, and which receives and integrates information from baroreceptors, chemoreceptors and higher centres in the brain. Baroreceptors are receptors sensitive to blood pressure and are found in the **carotid arteries/aorta**. A **rise** in blood pressure activates these receptors, which respond by increasing the activity of **parasympathetic** nerve fibres supplying the heart; this **slows the heart down** and returns the system towards normal. In addition to this, **sympathetic** nerve fibres supplying the blood vessels are **inhibited**, which leads to **vasodilation**, again returning the system towards normal (note that most blood vessels have little or no **parasympathetic** innervation).

On the other hand, if the blood pressure **falls**, baroreceptor activity is decreased, and this also triggers compensatory mechanisms. This time, **sympathetic** activity is increased and this leads to an **increase** in heart rate; in addition, cardiac contractile force is **increased**. The blood vessels respond with **vasoconstriction**; this is mainly due to **increased** activity in **sympathetic** fibres. These measures lead to a restoration of blood pressure towards normal.

In addition to the activity of the baroreceptors described above, chemoreceptors in the **carotid bodies/aorta** measure the pH of the blood. Increase in **carbon dioxide** content of the blood decreases pH and **stimulates** these receptors, leading to an **increase** in stroke volume and heart rate, and a general **vasoconstriction**; this **increases** blood pressure. Other control mechanisms include the renin–angiotensin system, which is involved in **long-term** regulation; activation **increases** blood volume, thereby **increasing** blood pressure.

64. Outwith the CNS: carotid bodies (carotid arteries) and aortic bodies (aorta); within the CNS: on the surface of the medulla oblongata.

65. Rising CO_2, falling O_2, falling pH.

66.

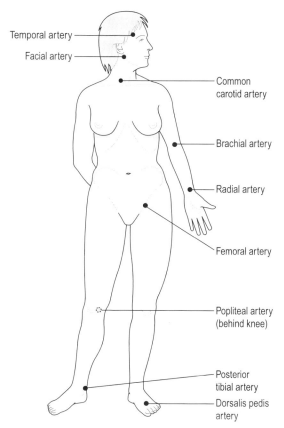

Figure 5.11

67. Blood leaving the right ventricle first enters the **pulmonary trunk**, which passes upwards close to the aorta and divides into the **right pulmonary**

arteries and the **left pulmonary arteries** at the level of the 5th thoracic vertebra. Each of these branches goes to the corresponding **lung** and enters these organs in the area called the **hilum or root**. Within the tissues, the vessels divide and subdivide, giving a network of many millions of tiny **capillaries**, across the walls of which gases exchange. Blood draining these structures then passes through veins of increasing diameter, which finally unite in the **pulmonary veins**, which carry the blood back to the **left** atrium of the heart.

68.

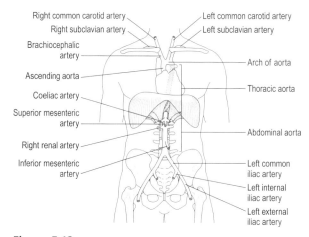

Figure 5.12

69. **Table 5.4** Features of some major arteries supplying the head and neck

Statement	Artery
Main supplier to the circulus arteriosus	Internal carotid artery
Supplies the superficial tissues of the head and neck	External carotid artery
Carotid sinuses occur where this artery bifurcates	Common carotid artery
Excepting the coronary arteries, this is the first artery to branch from the arch of the aorta	Brachiocephalic artery
One of the arteries forming the circulus arteriosus	Basilar artery
This artery can be felt as a pulse point just in front of and above the ear	Temporal artery

70.

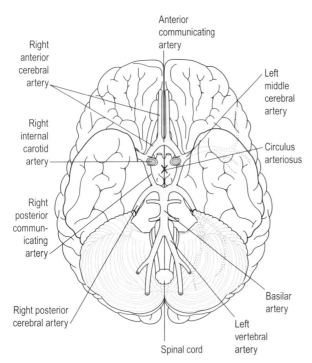

Figure 5.13

71.

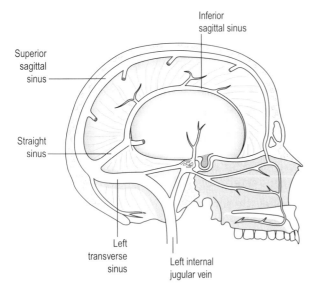

Figure 5.14

72.

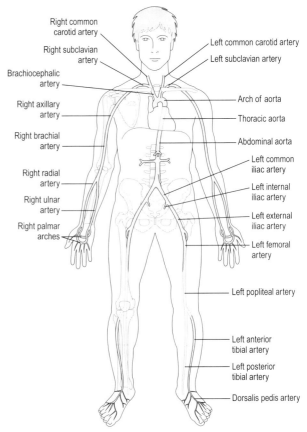

Figure 5.15

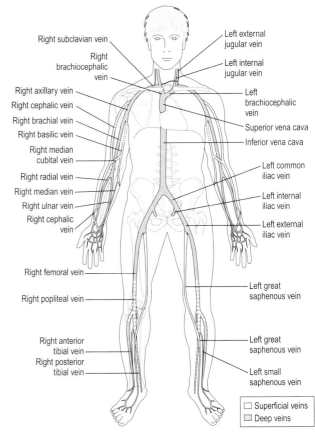

Figure 5.16

73.

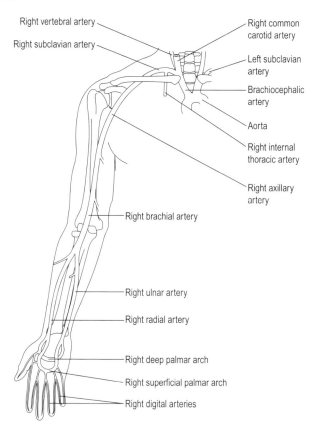

Right vertebral artery
Right subclavian artery
Right common carotid artery
Left subclavian artery
Brachiocephalic artery
Aorta
Right internal thoracic artery
Right axillary artery
Right brachial artery
Right ulnar artery
Right radial artery
Right deep palmar arch
Right superficial palmar arch
Right digital arteries

Figure 5.17

74. Aorta, common iliac artery, external iliac artery, femoral artery, popliteal artery, anterior tibial artery, dorsalis pedis artery, digital arteries, digital veins, dorsal venous arch, anterior tibial vein, popliteal vein, femoral vein, external iliac vein, common iliac vein, inferior vena cava.

75. Table 5.5 Features of some major vessels

Artery	P or U	Statement
Internal iliac artery	P	branches to supply the pelvic organs
External iliac artery	P	becomes the femoral artery
Intercostal artery	P	runs along the inferior border of each rib
Superior mesenteric artery	U	supplies the small intestine
Phrenic artery	P	supplies the diaphragm
Coeliac artery	U	divides into the hepatic, splenic and gastric arteries
Abdominal aorta	U	splits into the right and left common iliac arteries
Cystic artery	U	supplies the gall bladder
Hepatic portal vein	U	links the small intestine and the liver

76. and **77.**

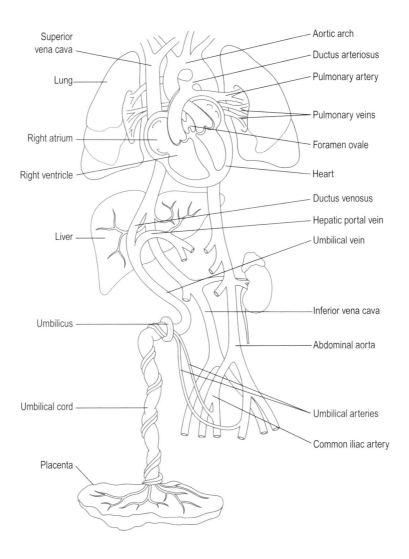

Superior vena cava

Lung

Right atrium

Right ventricle

Liver

Umbilicus

Umbilical cord

Placenta

Aortic arch

Ductus arteriosus

Pulmonary artery

Pulmonary veins

Foramen ovale

Heart

Ductus venosus

Hepatic portal vein

Umbilical vein

Inferior vena cava

Abdominal aorta

Umbilical arteries

Common iliac artery

Figure 5.18

78. Produces oestrogen and progesterone to maintain pregnancy; prevents harmful substances crossing into foetal circulation; allows exchange of nutrients and wastes between maternal and foetal circulation.

The lymphatic system

ANSWERS

1. and **2.**

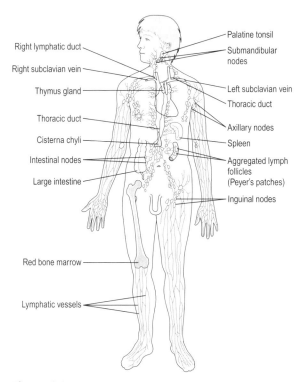

Right lymphatic duct
Right subclavian vein
Thymus gland
Thoracic duct
Cisterna chyli
Intestinal nodes
Large intestine

Palatine tonsil
Submandibular nodes
Left subclavian vein
Thoracic duct
Axillary nodes
Spleen
Aggregated lymph follicles (Peyer's patches)
Inguinal nodes

Red bone marrow

Lymphatic vessels

Figure 6.1

3. Although lymph and interstitial fluid are identical in composition, interstitial fluid bathes the cells and lymph is found in the lymphatic vessels.

4. b.   **5.** a., b., c.   **6.** d.   **7.** c.

8. The smallest lymphatic vessels are called **capillaries**. One significant difference between them and the smallest blood vessels is that they **originate in the tissues**; their function is to drain the lymph, containing **white blood cells**, away from the interstitial spaces. Most tissues have a network of these tiny vessels, but one notable exception is **bone tissue**. The individual tiny vessels join up to form larger ones, which now contain **three** layers of tissue in their walls, similar to veins in the cardiovascular system. The inner lining, the **endothelial** layer, covers the valves, which **regulate flow of lymph**. Unlike the cardiovascular system, there is no organ acting as a pump to push lymph

through the vessels, but forward pressure is applied to the lymph by various mechanisms, including **squeezing of the vessels by external structures like skeletal muscle/intrinsic contractility of the smooth muscle of lymphatic vessel walls**. As vessels progressively unite and become wider and wider, eventually they empty into the biggest lymph vessels of all, the **thoracic duct and the right lymphatic duct**. The first one of these drains the **right side of the body above the diaphragm**. The second drains the **lower part of the body and the upper left side above the diaphragm**.

9. and **10.**

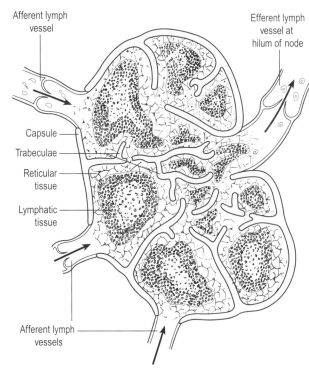

Afferent lymph vessel

Efferent lymph vessel at hilum of node

Capsule
Trabeculae
Reticular tissue
Lymphatic tissue

Afferent lymph vessels

Figure 6.2

11. Malignant cells; infected cells; microbes; inhaled particles; cell debris; worn out cells; damaged cells.

12. c.   **13.** a.   **14.** d.   **15.** b.

16. Literally, 'cell eating', the ingestion of unwanted or foreign cells or particles by the body's defence cells, with the intention of destroying or neutralizing them.

253

17. **Table 6.1** Characteristics of lymph nodes, spleen and thymus

Spleen	Thymus	Lymph node
Largest lymphatic organ	Maximum weight usually 30–40 g	Size from pin head to almond sized
Lies immediately below the diaphragm	Lies immediately behind the sternum	Distributed throughout lymphatic system
Stores blood	Secretes the hormone thymosin	Phagocytoses cellular debris
Oval in shape	Made up of two narrow lobes	Bean-shaped
Synthesizes red blood cells in the fetus	T-lymphocytes mature here	Site of multiplication of activated lymphocytes
Red blood cells destroyed here	At its maximum size at puberty	Filters lymph

18. c.   19. a.   20. c.   21. b.

The nervous system

ANSWERS

1. Brain, spinal cord.

2. and 3. See Figure 7.1.

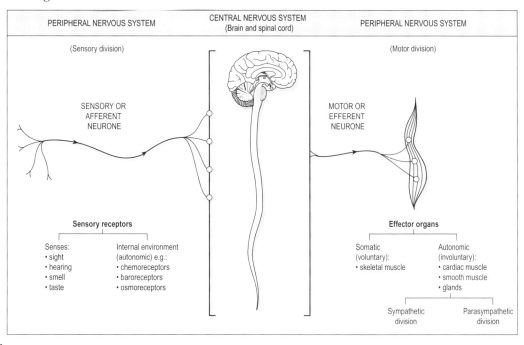

Figure 7.1

4. and 5. See Figure 7.2.

6. Myelinated neurones have nodes of Ranvier, non-myelinated neurones do not. One Schwann cell surrounds the axons of many non-myelinated neurones.

7. a.   8. d.   9. b.   10. a.   11. c.   12. c.

13. Transmission of the **action potential**, or impulse, is due to movement of **ions** across the nerve cell membrane. In the resting state the nerve cell membrane is **polarized** due to differences in the concentrations of ions across the plasma membrane. This means that there is a different electrical charge on each side of the membrane, which is called the resting **membrane potential**. At rest the charge outside the cell is **positive** and inside it is **negative**. The principal ions involved are **sodium** and **potassium**. In the resting state there is a continual tendency for these ions to diffuse down their **concentration gradients**. During the action potential, sodium ions flood **into** the neurone causing **depolarization**. This is followed by **repolarization** when potassium ions move **out of** the neurone. In myelinated neurones the insulating properties of the **myelin sheath** prevent the movement of ions across the membrane where this is present. In these neurones, impulses pass from one **node of Ranvier** to the next and transmission is called **saltatory conduction**. In unmyelinated fibres impulses are conducted by the process called **simple propagation**. Impulse conduction is faster when the mechanism of transmission is **saltatory conduction** than when it is **simple propagation**. The diameter of the neurone also affects the rate of impulse conduction: the **larger** the diameter, the faster the conduction.

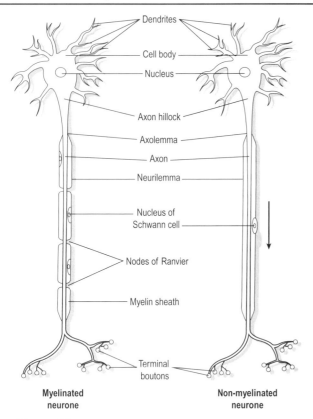

Figure 7.2

14. 15. and **16.**

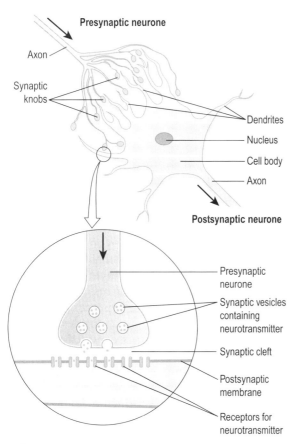

Figure 7.3

17. Acetylcholine.

18. The region where a nerve impulse passes from one neurone to another is called the synapse. The distal end of the presynaptic neurone breaks up into minute branches known as **synaptic knobs/terminal boutons**. These are in close proximity to the dendrites and cell bodies of the **postsynaptic neurone**. The space between them is the **synaptic cleft**. In the ends of the presynaptic neurones are spherical structures called **synaptic vesicles** containing chemicals known as the **neurotransmitter**. When the action potential depolarizes the presynaptic membrane, the chemicals in the membrane-bound packages are released into the synaptic cleft by the process of **exocytosis**. The chemicals released then move across the synaptic cleft by **diffusion**. They act on specific areas of the postsynaptic membrane called receptors causing **depolarization**.

19.

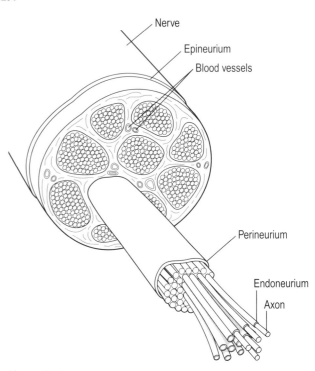

Figure 7.4

20. **Table 7.1** Characteristics of the meninges

	Dura mater	Arachnoid mater	Pia mater
Consists of two layers of dense fibrous tissue	✓		
Consists of fine connective tissue			✓
The epidural space lies above this layer	✓		
The subdural space lies between these two layers	✓	✓	
Surrounds the venous sinuses	✓		
The subarachnoid space separates these two layers		✓	✓
Forms the filum terminale			✓
CSF is found in the space between these two layers		✓	✓
Equivalent to the periosteum of other bones	✓		

21.

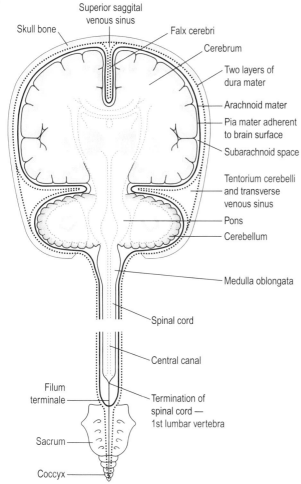

Figure 7.5

22. Protects the brain from potentially toxic substances and chemical variations in the blood.

23. Insertion of dyes for diagnostic purposes or for administration of drugs, e.g. analgesics or local anaesthetics.

24. Ventricles are shaded and components labelled **25.**

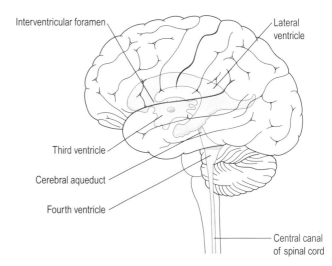

Figure 7.6

26. Cerebrospinal fluid (CSF).

27. a.   28. a., b., d.   29. d.   30. c.   31. d.   32. b.

33. Supports the brain in the cranial cavity, maintains uniform pressure around the brain and spinal cord, protects the brain and spinal cord by acting as a shock absorber between the brain and cranial bones, keeps the brain and spinal cord moist and may allow exchange of substances between CSF and nerve cells.

34. and 35.

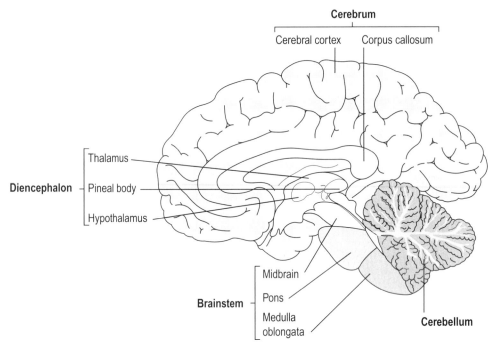

Figure 7.7

36. Circulus arteriosus (circle of Willis).

37. This is the largest part of the brain and is divided into left and right **cerebral hemispheres**. Deep inside, the two parts are connected by the **corpus callosum**, which consists of **white** matter. The superficial layer of the cerebrum is known as the **cerebral cortex** and consists of nerve **cell bodies** or **grey** matter. The deeper layer consists of nerve **fibres** and is **white** in colour. The cerebral cortex has many furrows and folds that vary in depth. The exposed areas are the convolutions or **gyri** and they are separated by **sulci**, also known as **fissures**. These convolutions increase the **surface area** of the cerebrum.

38.

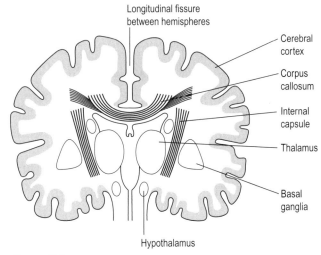

Figure 7.8

39. Mental activities, e.g. memory, learning, reasoning; sensory perception; initiation and control of skeletal muscle contraction.

40.

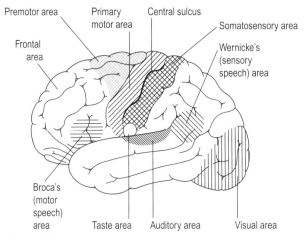

Premotor area   Primary motor area   Central sulcus
Frontal area   Somatosensory area
Wernicke's (sensory speech) area
Broca's (motor speech) area   Taste area   Auditory area   Visual area

Figure 7.9

41. **a.** Cell bodies; **b.** axons and dendrites.

42. The primary motor area lies in the **frontal** lobe immediately anterior to the **central** sulcus. The cell bodies are **pyramid-shaped** and stimulation leads to contraction of **skeletal** muscle. Their nerve fibres pass downwards through the **internal capsule** to the **medulla** where they cross to the opposite side then descend in the **spinal cord**. These neurones are the upper motor neurones. They synapse with the lower motor neurones in the **spinal cord** and lower motor neurones terminate at a **neuromuscular junction**. This means that the motor area of the right hemisphere controls skeletal muscle movement on **the left side** of the body.

In the motor area of the cerebrum, body areas are represented **upside down** and the proportion of the cerebral cortex that represents a particular part of the body reflects its **complexity of movement**.

Broca's area lies in the **frontal** lobe and controls the movements needed for **speech**. The right hemisphere is dominant in **left-handed** people.

The frontal area is situated in the **frontal** lobe and is thought to be involved in one's **character**.

43. **a.** Sensory speech area; **b.** gustatory area; **c.** auditory area; **d.** olfactory area; **e.** visual area; **f.** visual area; **g.** auditory area; **h.** gustatory area and olfactory area; **i.** visual area; **j.** auditory area.

44. b., c., d.   45. d.   46. a.   47. b., d.   48. a.

49. a., b.   50. a., b., c., d.   51. a., c.

52. a. Selective awareness, which blocks or transmits sensory information to the cerebral cortex, e.g. a crying child; b. co-ordination of voluntary movement, posture and balance.

53. 45 cm.   54. 1st lumbar vertebra.

55. Insertion of a cannula into the subarachnoid space below the spinal cord (i.e. below the 2nd lumbar vertebra) to measure CSF pressure and/or obtain a sample of CSF.

56. a. Origin – spinal cord, destination – thalamus; b. origin – cerebral cortex, destination – spinal cord.

57. **Table 7.2** Characteristics of the motor and sensory pathways of the spinal cord

	Motor pathways	Sensory pathways
Impulses travel towards the brain		✓
The extrapyramidal tracts are an example of these	✓	
Consist of two neurones	✓	
Contain afferent tracts		✓
Their fibres pass through the internal capsule	✓	
Impulses from proprioceptors travel via these pathways		✓
Are involved in fine movements	✓	
Are involved in movement of skeletal muscles	✓	
Impulses follow activation of receptors in the skin		✓
Impulses travel away from the brain	✓	
May consist of either two or three neurones		✓

58., 59. and 60.

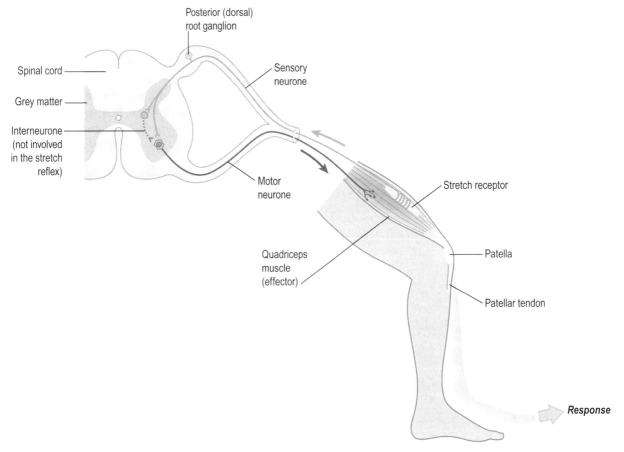

Figure 7.10

61. The connector neurone/interneurone.

62. Within the peripheral nervous system there are **31** pairs of spinal nerves and **12** pairs of cranial nerves. These nerves are composed of either **sensory** nerve fibres conveying afferent impulses to **the brain** from **sensory** organs, or **motor** nerve fibres that transmit efferent impulses from **the brain** to **effector** organs. Some nerves, known as **mixed** nerves, contain both types of fibres.

63. It is a site where spinal nerves are regrouped before going on to their destination, meaning that damage to one spinal nerve does not cause loss of function of an area.

64.

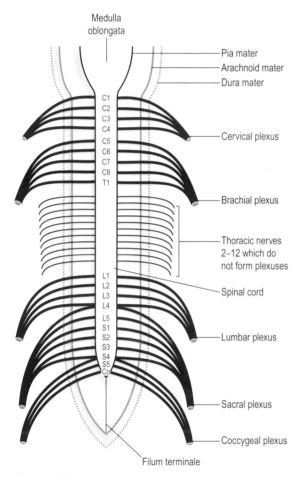

Figure 7.11

65. and **66.**

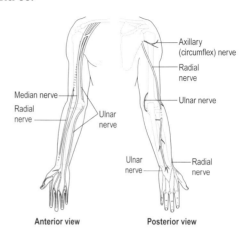

Figure 7.12

67. and **68.**

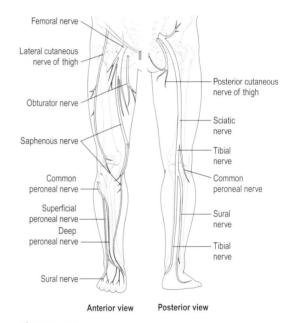

Figure 7.13

69. **Table 7.3** Nerves supply to muscles

Muscle supplied	Name of nerve	Plexus of origin
Intercostal muscles	Intercostal	N/A
Diaphragm	Phrenic	Cervical
Quadriceps	Femoral	Lumbar
Hamstrings	Sciatic	Sacral
External anal sphincter	Pudendal	Sacral
External urethral sphincter	Pudendal	Sacral

70., 71. and **72.**

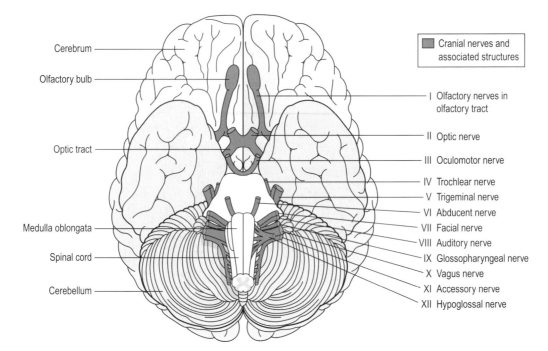

Figure 7.14

73. and **74. Table 7.4** The cranial nerves and their functions

Number	Name	Function	Type
I	Olfactory	Sense of smell	Sensory
II	Optic	Sense of sight, balance	Sensory
III	Oculomotor	Moving the eyeball, focusing, regulating the size of the pupil	Motor
IV	Trochlear	Movement of the eyeball	Motor
V	Trigeminal	Chewing, sensation from the face	Mixed
VI	Abducent	Movement of the eye	Motor
VII	Facial	Sense of taste, movements of facial expression	Mixed
VIII	Vestibulocochlear	Maintaining balance, sense of hearing	Sensory
IX	Glossopharyngeal	Secretion of saliva, sense of taste, movement of pharynx	Mixed
X	Vagus	Movement and secretion in GI tract, heart rate	Mixed
XI	Accessory	Movement of the head, shoulders and larynx	Motor
XII	Hypoglossal	Movement of the tongue	Motor

75. c.   **76.** d.   **77.** a.   **78.** b.   **79.** b.   **80.** c.   **82.** Sympathetic, parasympathetic.

81. Smooth muscle, cardiac muscle, glands.   **83.** a. F; b. F; c. T; d. F; e. T; f. F; g. F; h. T; i. F; j. T; k. F.

84. Represented by the dotted lines on Figure 7.15.

85., 86. and **87.** See Figure 7.15.

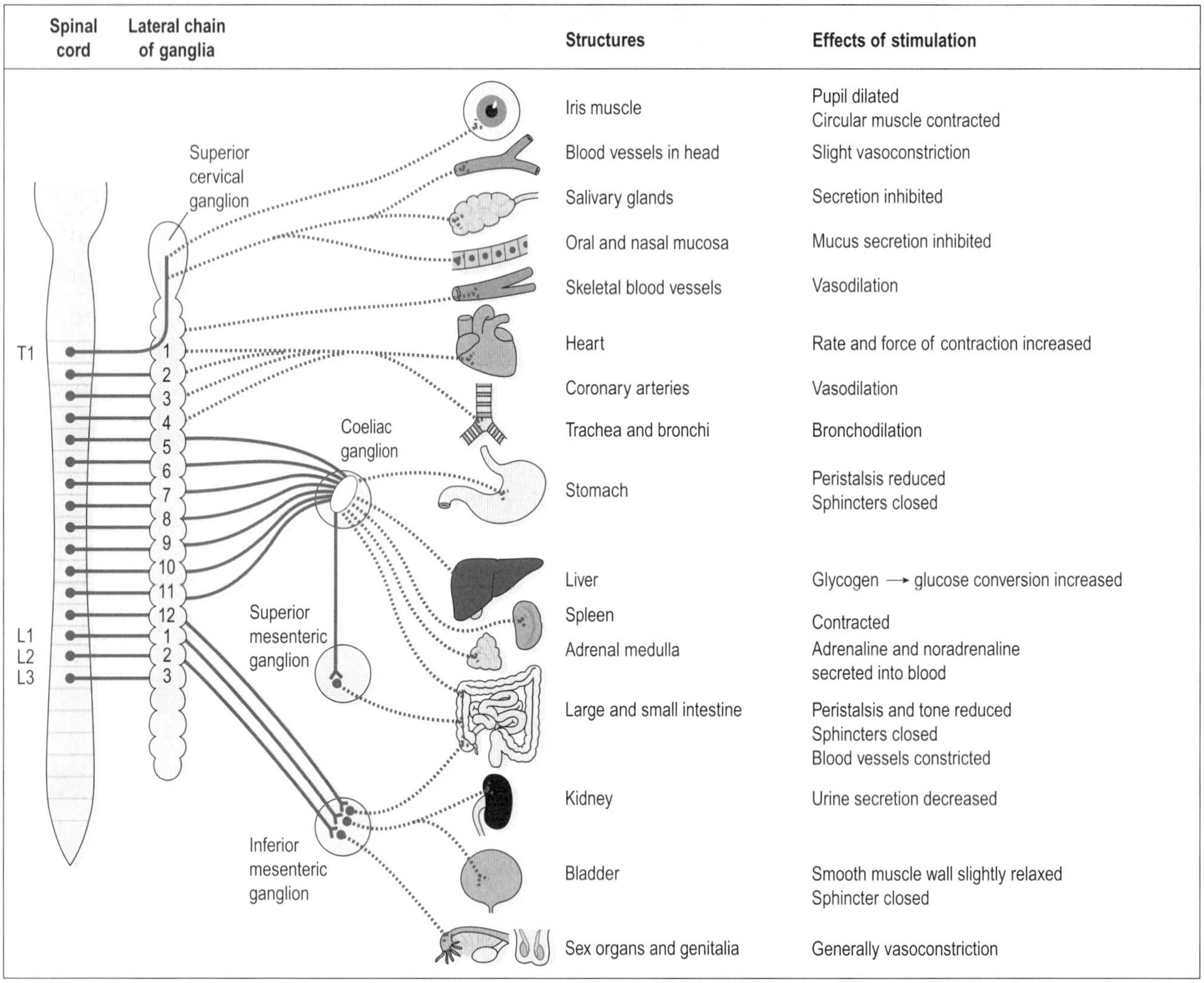

Spinal cord	Lateral chain of ganglia	Structures	Effects of stimulation
	Superior cervical ganglion	Iris muscle	Pupil dilated Circular muscle contracted
		Blood vessels in head	Slight vasoconstriction
		Salivary glands	Secretion inhibited
		Oral and nasal mucosa	Mucus secretion inhibited
		Skeletal blood vessels	Vasodilation
T1	1, 2, 3, 4, 5, 6, 7, 8, 9, 10, 11, 12	Heart	Rate and force of contraction increased
	Coeliac ganglion	Coronary arteries	Vasodilation
		Trachea and bronchi	Bronchodilation
		Stomach	Peristalsis reduced Sphincters closed
	Superior mesenteric ganglion	Liver	Glycogen $\rightarrow$ glucose conversion increased
L1, L2, L3	1, 2, 3	Spleen	Contracted
		Adrenal medulla	Adrenaline and noradrenaline secreted into blood
		Large and small intestine	Peristalsis and tone reduced Sphincters closed Blood vessels constricted
	Inferior mesenteric ganglion	Kidney	Urine secretion decreased
		Bladder	Smooth muscle wall slightly relaxed Sphincter closed
		Sex organs and genitalia	Generally vasoconstriction

Figure 7.15

88. Pain from an internal organ perceived to originate elsewhere in the body, e.g. angina.

89. Sensory fibres from the affected organ enter the same segment of the spinal cord as sensory fibres from the area of perceived pain and the brain perceives them as coming from the latter source.

90. Represented by the dotted lines on Figure 7.16.

91. In other organs, the cell bodies of parasympathetic postganglionic neurones lie in the wall of the structure supplied and therefore the postganglionic neurone is very small.

92. and **93.**

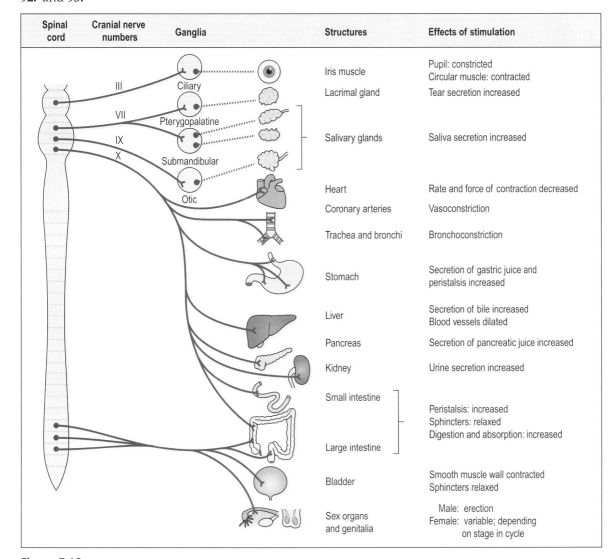

Spinal cord	Cranial nerve numbers	Ganglia	Structures	Effects of stimulation
	III	Ciliary	Iris muscle	Pupil: constricted Circular muscle: contracted
	VII	Pterygopalatine	Lacrimal gland	Tear secretion increased
	IX	Submandibular	Salivary glands	Saliva secretion increased
	X	Otic	Heart	Rate and force of contraction decreased
			Coronary arteries	Vasoconstriction
			Trachea and bronchi	Bronchoconstriction
			Stomach	Secretion of gastric juice and peristalsis increased
			Liver	Secretion of bile increased Blood vessels dilated
			Pancreas	Secretion of pancreatic juice increased
			Kidney	Urine secretion increased
			Small intestine	Peristalsis: increased Sphincters: relaxed Digestion and absorption: increased
			Large intestine	
			Bladder	Smooth muscle wall contracted Sphincters relaxed
			Sex organs and genitalia	Male: erection Female: variable; depending on stage in cycle

Figure 7.16

ANSWERS

1. and **2.**

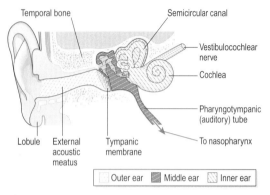

Figure 8.1

3. Earwax (wax).

4.

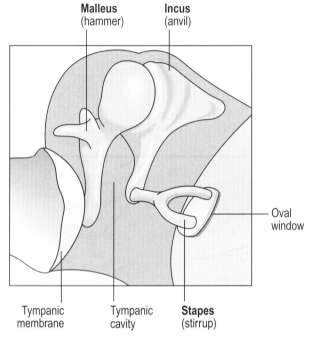

Figure 8.2

5. and **6.** See Figure 8.3.

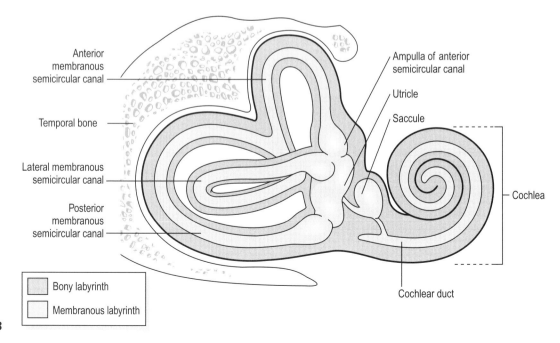

Figure 8.3

7. Vestibule.

8. and **9.**

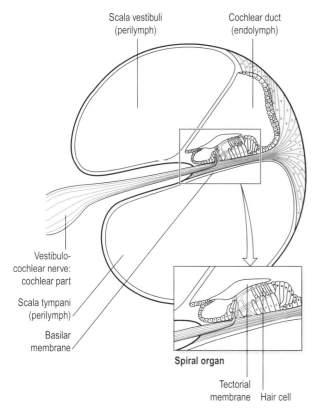

Figure 8.4

10. a. T; b. F; c. T; d. F; e. F; f. F.

11. A sound produces **waves/vibrations** in the air. The auricle **collects** and **directs** them along the **auditory canal** to the **tympanic membrane**. The vibrations are **transmitted** and **amplified** through the middle ear by movement of the **(auditory) ossicles**. At its medial end, movement of the **stapes** in the **oval window** sets up fluid waves in the **perilymph** of the scala vestibuli. Most of this pressure is transmitted into the **cochlear duct** resulting in a corresponding fluid wave in the **endolymph**. This stimulates the auditory receptors in the **hair** cells in the organ of hearing, **the spiral organ (of Corti)**. Stimulation of the auditory receptors results in the generation of **nerve impulses** that travel to the brain along the **cochlear/auditory** part of the **vestibulocochlear** nerve. The fluid wave is extinguished by vibration of the membrane of the **round** window.

12., 13. and **14.**

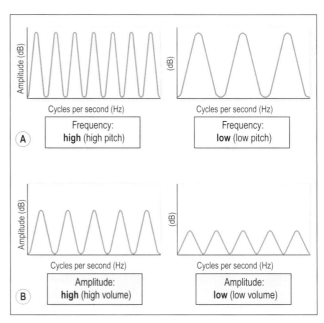

Figure 8.5

15. The organs involved with balance are found in the **inner** ear. They are the three **semicircular** canals, one in each plane of space, and the vestibule, which comprises two parts, the **saccule** and the utricle. The canals, like the cochlea, are composed of an outer bony wall and inner membranous ducts. The membranous ducts contain **endolymph** and are separated from the bony wall by **perilymph**. They have dilated portions near the vestibule called ampullae containing hair cells with sensory nerve endings between them. Any change in the position of the head causes movement in the endolymph and perilymph. This causes stimulation of the hair cells and nerve impulses are generated. These travel in the vestibular part of the vestibulocochlear nerve to the **cerebellum** via the **vestibular** nucleus. Perception of body position occurs because the cerebrum co-ordinates impulses from the eyes and proprioceptors in addition to those from the cerebellum.

16. and **17.**

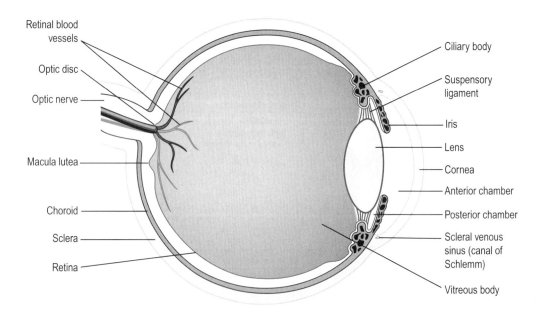

Figure 8.6

18.

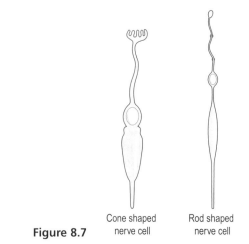

Cone shaped   Rod shaped
nerve cell   nerve cell

Figure 8.7

19. The anterior segment of the eye is incompletely divided into the **anterior** and **posterior** chambers by the **iris**. Both chambers contain **aqueous fluid** secreted into the **posterior** chamber by the **ciliary glands**. It circulates in front of the **lens** and through the **pupil** into the **anterior** chamber and returns to the circulation through the **scleral venous sinus**. As there is continuous production and drainage, the intraocular pressure remains fairly constant. The structures in the front of the eye including the **cornea** and the **lens** are supplied with nutrients by the **aqueous fluid**. The posterior segment of the eye lies behind the **lens** and contains the **vitreous body**. It has the consistency of **jelly** and provides sufficient intraocular pressure to keep the eyeball from collapsing.

20. The retina lines the **eyeball**. Near the centre is the **macula lutea** or yellow spot, consisting only

of **cone**-shaped cells. The small area of the retina where the optic nerve leaves is the **optic disc** or **blind spot**.

21.

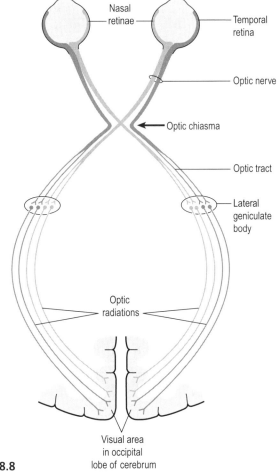

Figure 8.8

22. d.   **23.** a.   **24.** c.   **25.** c.   **26.** b.   **27.** c.   **28.** a.   **29.** c., d.

30.

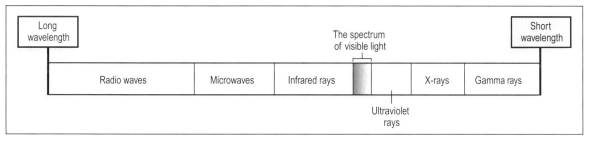

Figure 8.9

31., 32. and **33.**

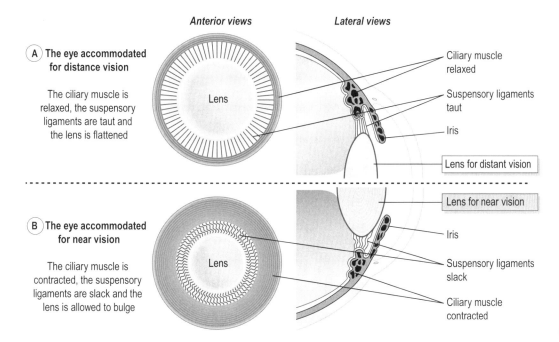

Figure 8.10

34. About 6 metres.

35. The amount of light entering the eye is controlled by the **size** of the pupils. In a bright light they are **constricted** and in darkness they are **dilated**. The iris consists of two layers of smooth muscle – contraction of the circular fibres causes **constriction** of the pupil while contraction of the radiating fibres causes **dilation**. The autonomic nervous system controls the size of the pupil – sympathetic stimulation causes **dilation** while parasympathetic stimulation causes **constriction** of the pupil.

36. Constriction of the pupils, convergence of the eyeballs, changing the power of the lens.

37. In bright light rhodopsin present in rods is completely degraded. On moving into a darkened area where the light intensity is insufficient to stimulate the cones there is temporary visual impairment until normal sight returns when the rhodopsin has been regenerated.

38. **Table 8.1** Actions of the extrinsic muscles of the eye

Extrinsic muscle	Action
Medial rectus	Rotates the eyeball inwards
Lateral rectus	Rotates the eyeball outwards
Superior rectus	Rotates the eyeball upwards
Inferior rectus	Rotates the eyeball downwards
Superior oblique	Rotates the eyeball downwards and outwards
Inferior oblique	Rotates the eyeball upwards and outwards

39. and **40.**

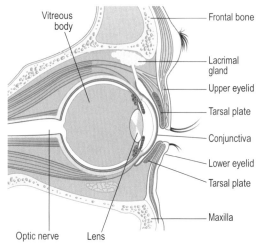

Figure 8.11

41. Lacrimal glands.

42. Water, mineral salts, antibodies, lysozyme.

43. Washing away irritants; the bactericidal enzyme lysozyme prevents infection; the oily secretion from the tarsal glands delays evaporation and prevents drying of the conjunctiva; nourishment of the cornea.

44. All odorous materials give off **volatile** molecules that are carried into the nose in the inhaled air and stimulate the olfactory **chemoreceptors**. When currents of air are carried to the **roof of the nasal cavity** the smell receptors are stimulated, setting up impulses in the olfactory nerve endings. These pass through the cribriform plate of the **ethmoid bone** to the olfactory bulb. Nerve fibres that leave the olfactory bulb form the olfactory tract. This passes posteriorly to the olfactory lobe of the **cerebrum or cerebral cortex** where the impulses are interpreted and odour perceived.

45. Absence of the sense of smell.

46. Perception of a particular smell decreases and stops after a few minutes of exposure.

47. Taste buds contain sensory receptors called **chemoreceptors**. They are situated in the papillae of the **tongue** and in the epithelia of the tongue, **soft palate, pharynx** and **epiglottis**. Some of the taste buds have hair-like **cilia** on their free border projecting towards tiny pores in the epithelium. Sensory receptors are stimulated by chemicals dissolved in **saliva** and **nerve impulses** are generated when stimulation occurs. These are conducted to the brain where taste is perceived by the **taste/gustatory** area in the **parietal** lobe of the cerebral cortex.

48. Stimulates salivation and secretion of gastric juice.

49. Can initiate gagging or vomiting reflexes.

50. and **51.**

52. Sweet, salt, sour, bitter.

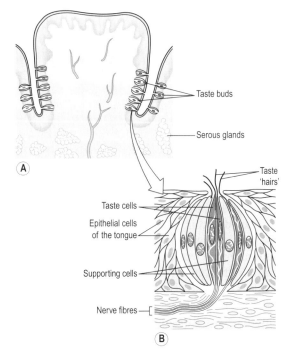

Figure 8.12

53. An abnormal curvature of part of the cornea or lens prevents focusing on the retina, resulting in blurred vision.

54. Nearsightedness – the eyeball is too short, resulting in focusing of near images behind the retina.

55. Farsightedness – the eyeball is too long, causing a far image to be focused in front of the retina.

56.

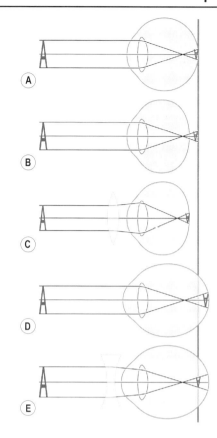

Figure 8.13

ANSWERS

1.

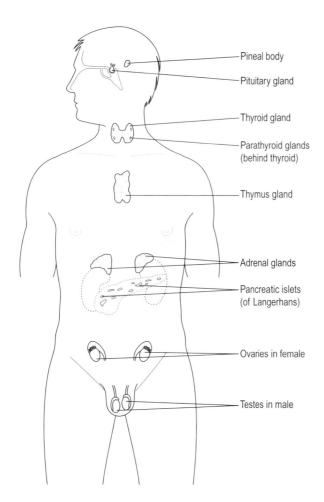

- Pineal body
- Pituitary gland
- Thyroid gland
- Parathyroid glands (behind thyroid)
- Thymus gland
- Adrenal glands
- Pancreatic islets (of Langerhans)
- Ovaries in female
- Testes in male

Figure 9.1

2. Four.

3. There are two embedded in the posterior surface of each lobe of the thyroid gland.

4. A hormone is formed by a gland that **secretes** it into **the bloodstream** which transports it to its target organ. When it arrives at its site of action, it binds to specific sites known as **receptors** to bring about its effect. Many actions of the endocrine system are concerned with maintaining homeostasis of the **internal** environment. This often takes place in co-ordination with the **nervous** system. The effects of the endocrine system are usually **slower** and **more** precise than the other system. Hormones may be either lipid-based, e.g. **steroids**, or peptides, which are **water** soluble, e.g. **insulin**.

5. **Table 9.1** Anatomy of the pituitary gland

Connects the pituitary gland to the hypothalamus	Pituitary stalk
Composed of glandular tissue	Anterior lobe of the pituitary
Composed of nervous tissue	Posterior lobe of the pituitary
Part of the pituitary whose function is unknown in humans	Intermediate lobe of the pituitary
Situated superiorly to the pituitary gland	Hypothalamus
A supporting cell of the posterior pituitary	Pituicyte

6. **Table 9.2** Hormones of the hypothalamus, anterior pituitary and their target tissues

Hypothalamus	Anterior pituitary	Target gland or tissue
GHRH – growth hormone releasing hormone	GH	Most tissues, many organs
GHRIH – growth hormone release inhibiting hormone	GH inhibition TSH inhibition	Thyroid gland, pancreatic islets, most tissues
TRH – thyrotrophin releasing hormone	TSH	Thyroid gland
CRH – corticotrophin releasing hormone	ACTH	Adrenal cortex
PRH – prolactin releasing hormone	PRL	Breast
PIH – prolactin inhibiting hormone	PIH	Breast
LHRH – luteinizing releasing hormone, also known as GnRH – gonadotrophin releasing hormone	FSH LH	Ovaries and testes Ovaries and testes

7. **Table 9.3** Summary of the hormones secreted by the anterior pituitary gland

Hormone	Abbreviation	Function
Growth hormone	GH	Regulates metabolism, promotes tissue growth – especially bone
Thyroid stimulating hormone	TSH	Stimulates growth and activity of the thyroid gland
Adrenocorticotrophic hormone	ACTH	Stimulates the adrenal glands to secrete glucocorticoids
Prolactin	PRL	Stimulates milk production in the mammary glands
Follicle stimulating hormone	FSH	Males: stimulates production of sperm in the testes Females: stimulates secretion of oestrogen in the ovaries, maturation of ovarian follicles, ovulation
Luteinizing hormone	LH	Males: stimulates secretion of testosterone in the testes Females: stimulates secretion of progesterone by the corpus luteum

8.

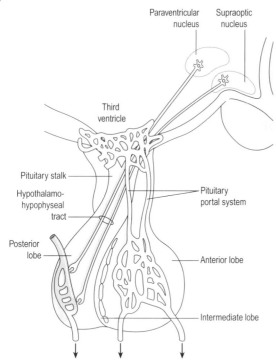

Paraventricular nucleus

Supraoptic nucleus

Third ventricle

Pituitary stalk

Hypothalamo-hypophyseal tract

Posterior lobe

Pituitary portal system

Anterior lobe

Intermediate lobe

Figure 9.2

9. Antidiuretic hormone (ADH) and oxytocin.

10. Adenohypophysis.

11. Neurohypophysis.

12. and **13.**

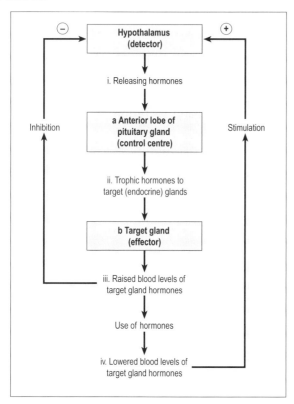

Hypothalamus (detector)

−   +

i. Releasing hormones

Inhibition

Stimulation

a Anterior lobe of pituitary gland (control centre)

ii. Trophic hormones to target (endocrine) glands

b Target gland (effector)

iii. Raised blood levels of target gland hormones

Use of hormones

iv. Lowered blood levels of target gland hormones

Figure 9.3

14. a., b., c.   **15.** d.   **16.** a.   **17.** b.   **18.** c.

19. An increase in the rate of urine production is called **diuresis**. ADH is secreted by the **posterior** pituitary gland; its main effect is to **decrease** urine output. It does this by **increasing** the permeability of the **distal** convoluted tubules and the **collecting** ducts in the nephrons of the kidneys to water. As a result, more **water** is reabsorbed from the filtrate. Secretion of ADH occurs in response to increasing **osmotic pressure** of the blood, which is detected by **osmo**-receptors in the hypothalamus. Situations where this takes place include **dehydration** and **haemorrhage (shock)** – more water is reabsorbed decreasing the blood **osmotic pressure**. In more serious situations, ADH also causes **contraction** of smooth muscle, which results in **vasoconstriction** in small arteries. This has a pressor effect, increasing **systemic blood** pressure, which reflects the alternative name of this hormone, **vasopressin**.

20. Oxytocin stimulates two target tissues before and after childbirth. These are **uterine smooth muscle** and **myoepithelial cells** of the lactating breast. During childbirth, also known as **parturition**, increasing amounts of oxytocin are released in response to increasing **stimulation** of sensory **stretch receptors** in the **(uterine) cervix** by the baby's head. Sensory impulses are generated and travel to the **hypothalamus** stimulating the **posterior pituitary** to secrete more oxytocin. This **stimulates** the uterus to contract more forcefully moving the baby's head further downwards through the uterine cervix and vagina. The mechanism stops shortly after the baby has been born. This is an example of a **positive** feedback mechanism. After birth oxytocin stimulates **lactation**.

21. **Table 9.4** Features of the thyroid gland

The thyroid gland is surrounded by this structure	Capsule
Joins the two thyroid lobes together	Isthmus
Lie against the posterior surface of the thyroid gland	Parathyroid glands
Secrete the hormone calcitonin	Parafollicular cells
Constituent of T_3 and T_4	Iodine
Secreted by the hypothalamus	TRH
Thyroxine	T_4
Precursor of T_3 and T_4	Thyroglobulin
Secreted by the anterior pituitary	TSH
The nerves close to the thyroid gland	Recurrent laryngeal

24. **Table 9.5** Effects of abnormal secretion of thyroid hormones

Body function affected	Hypersecretion of T_3 and T_4	Hyposecretion of T_3 and T_4
Metabolic rate	Increased	Decreased
Weight	Loss	Gain
Appetite	Good	Poor, anorexia
Mental state	Anxious, excitable, restless	Depressed, lethargic, mentally slow
Scalp	Hair loss	Brittle hair
Heart	Tachycardia, palpitations, atrial fibrillation	Bradycardia
Skin	Warm and sweaty	Dry and cold
Faeces	Loose – diarrhoea	Dry – constipation
Eyes	Exophthalmos	None

22.

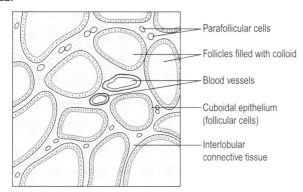

Figure 9.4

23. a. T; b. F; c. F; d. F.

25.

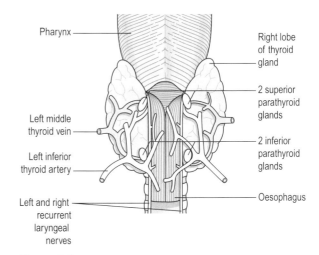

Figure 9.5

26. Parathyroid hormone (PTH).

27. Calatonin.

28. The parathyroid glands secrete parathyroid hormone (PTH) and blood calcium levels regulate its secretion. When they **fall**, secretion of PTH is increased and vice versa. The main function of PTH is to **increase** the blood calcium level. This is achieved by **increasing** the amount of calcium absorbed from the small intestine and reabsorbed from the renal tubules. If these sources do not provide sufficient levels then PTH stimulates **osteoclasts** (bone destroying cells) and calcium is released into the blood from **bones**. Normal blood calcium levels are needed for muscle **contraction**, blood clotting and nerve impulse transmission.

29. **Table 9.6** Features of the adrenal glands

Is essential for life	Cortex
Inner part of the adrenal gland	Medulla
Veins that drain the adrenal glands	Suprarenal
The organs immediately inferior to the adrenal glands	Kidneys
Male sex hormones	Androgens
The lipid that forms the basic structure of adrenocorticoids	Cholesterol
A mineralocorticoid hormone	Aldosterone
A glucocorticoid hormone	Hydrocortisone

30. and **31.** See Figure 9.6.

32. a. alpha (α); b. beta (β); c. delta (δ).

33. Growth hormone release inhibiting hormone (GHRIH).

34. a. T; b. F; c. T; d. F; e. F; f. T; g. F; h. T.

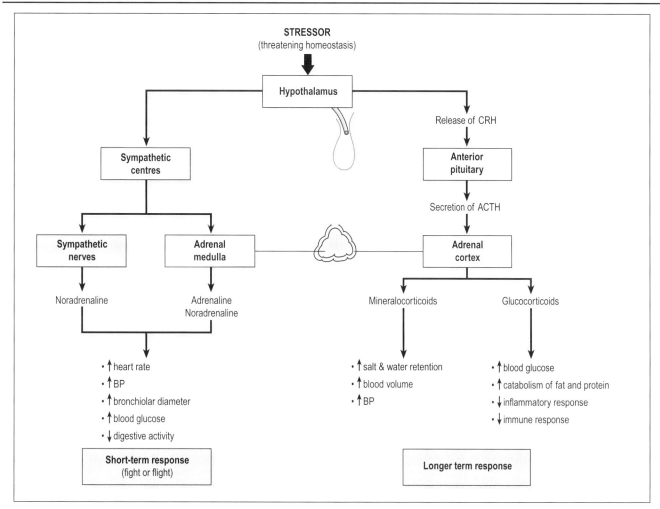

Figure 9.6

35. and **36. Table 9.7** The effect of insulin and glucagon on metabolic processes

Metabolic pathway	Effect of pathway on metabolism	Stimulated by insulin or glucagon?
Gluconeogenesis	Formation of new sugar from, e.g., protein	Glucagon
Lipogenesis	Promoting synthesis of fatty acids and storage of fat	Insulin
Glycogenesis	Increasing conversion of glucose to glycogen	Insulin
Glycogenolysis	Conversion of glycogen to glucose	Glucagon
Lipolysis	Breakdown of triglycerides to fatty acids	Insulin

37. a.   **38.** c.   **39.** a., b., c.   **40.** a., d.   **41.** b., c.

ANSWERS

1. and **2.**

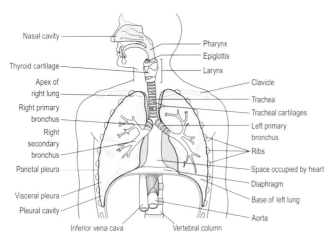

Figure 10.1

3. a. Nasal conchae, b. larynx, c. thyroid cartilage, d. epiglottis, e. soft palate, f. tonsils, g. auditory tube.

4.

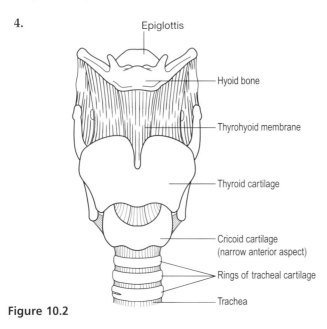

Figure 10.2

5. The upper respiratory passages carry air in and out of the respiratory system, but they have other functions too. The cells of their mucous membrane have **cilia**, tiny hair-like structures that **beat** in a wave-like motion towards the **mouth**. They carry mucus, which has been made by the **goblet** cells in the epithelial layer, and which traps **dirt** and **dust** on its sticky surface. The air is therefore **cleaned** by these mechanisms before it gets into the lungs. As the air passes through the nasal cavity, it is also **warmed** and **moistened** as it passes over the nasal **conchae**, bony projections covered in mucous membrane. The nasal cavity also contains **hair**, which is covered in **mucus**, and acts as a coarse filter for the air passing through. Immune tissue is present in patches called **tonsils**, which make **antibodies** and therefore protect against inhaled antigens. Not only air passes through the pharynx, but also **food** and **drink**, and the tracheal opening is barricaded against these by the **epiglottis**.

6.

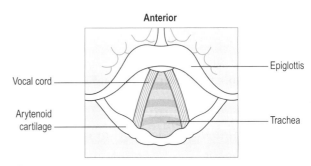

Figure 10.3

7. d.   **8.** a.   **9.** a.   **10.** c.

11. a. The vocal cords in the larynx are made of bands of <u>mucous membrane</u> stretched across the laryngeal lumen. b. Increasing the speed of vibration of the vocal cords increases the <u>volume</u> of sound produced. c. True. d. At rest, the vocal cords are adducted (<u>closed</u>). e. Sound is produced by air passing through the larynx on its way <u>from</u> the lungs, vibrating the vocal cords as it does so.

12.

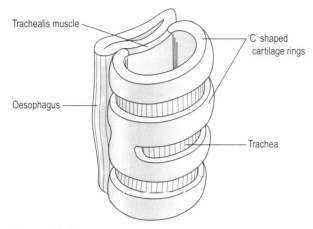

Figure 10.4

13. C-shaped, with the opening at the back (i.e. lying against the oesophagus).

14. To permit expansion of the oesophagus when swallowing, without obstructing the trachea.

15.

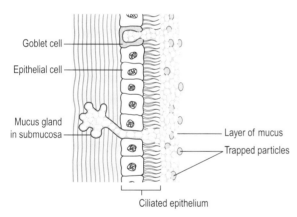

Figure 10.5

16. Cell A: Ciliated epithelial cell; protects and lines; cilia clear mucus and inhaled particles from the airways. Cell B: Goblet cell; produces sticky mucus that traps inhaled particles.

17. a. Mucus is produced in the upper respiratory tract because this an efficient way of removing dust and dirt from inhaled air. b. Cilia are present in the upper respiratory tract because mucus needs to be swept away from the lungs. c. Cartilage is present in the upper respiratory tract because the airways have to be kept open at all times. d. Elastic tissue is present in the upper respiratory tract because the passageway has to be flexible to allow head and neck movement.

18. Ciliated respiratory epithelium lines the **upper respiratory tract and wider airways only**, and its job is to keep the lungs clean. Cartilage rings support the airway walls; as the airways progressively divide and their diameter decreases, the amount of cartilage present **also decreases**. The smallest airways are called respiratory bronchioles, **and some** gas exchange takes place across their walls. The airways terminate in clusters of microscopic pouches called alveoli; it is here that most gas exchange takes place. The alveolar walls are only one cell thick and contain **alveolar cells**, which make surfactant to keep the alveoli from collapsing. Gas exchange occurring across the alveolar walls is called **external** respiration.

19. and **20.**

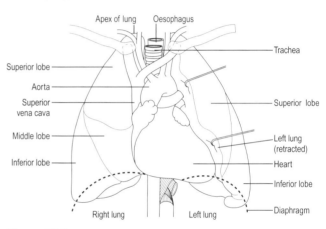

Figure 10.6

21. Mediastinum.

22.

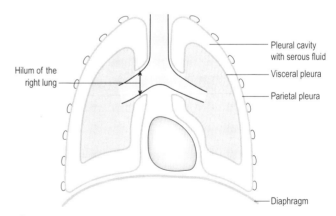

Figure 10.7

23. b, f.   **24.** c.   **25.** c.   **26.** b.   **27.** d.

28.

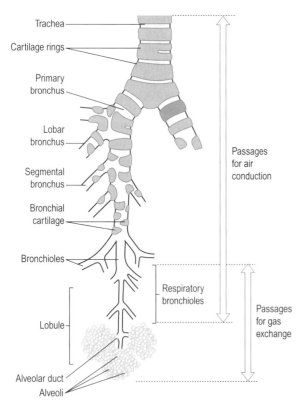

Figure 10.8

29. 150 million.

30. The elastic framework of the lung pulls on them and keeps them open.

31.

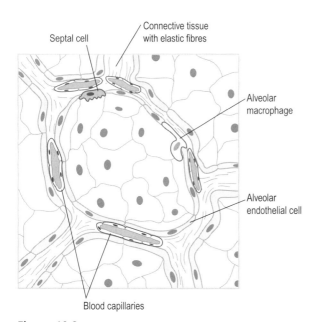

Figure 10.9

32. Cell A: surfactant, septal cell.
Cell B: phagocytosis; macrophage.

33. Left atrium, left ventricle, aorta, body tissues, right atrium, right ventricle, pulmonary artery, lungs, pulmonary vein.

34. Just before inspiration commences, the diaphragm is **relaxed**; this occurs in the pause between breaths in normal quiet breathing. Inspiration commences. The ribcage moves **upwards** and **outwards** owing to contraction of the **external intercostal muscles**. The diaphragm **contracts** and moves **downwards**. This **increases** the volume of the thoracic cavity, and **decreases** the pressure. Because of these changes, air moves **into** the lungs, and the lungs **inflate**. Inspiration has taken place.

Unlike inspiration, expiration is usually a **passive** process because it requires no **muscular effort**. So, following the end of inspiration, the diaphragm **relaxes** and moves back into its resting position. The ribcage moves **downwards** and **inwards**, because the **external intercostal muscles** have relaxed. This **decreases** the volume of the thoracic cavity, and so **increases** the pressure within it. Air therefore now moves **out of** the lungs and they **deflate**. There is now a rest period before the next cycle begins.

35. a. C; b. C/E; c. N; d. E.

36. b.

37.

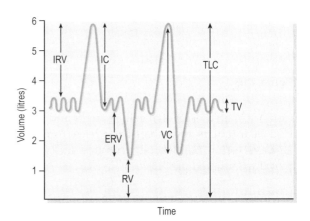

Figure 10.10

a. Tidal volume; b. vital capacity; c. inspiratory capacity; d. residual volume; e. inspiratory reserve volume; f. expiratory reserve volume.

38. IC is 3000 ml; IRV is 2500 ml.

39. 4800 ml.

40. 3770 ml.

41. Expiratory reserve volume.

42. Both residual volume and vital capacity are fixed measures, determined by individual anatomical and physiological constraints, and are unaffected by exercise.

43. Exchange of gases in the lung; oxygen leaves the alveoli and enters the blood, and carbon dioxide leaves the blood and enters the lung, both gases moving down their pressure gradients.

44. Exchange of gases in the tissues; oxygen leaves the blood and enters the tissues, and carbon dioxide leaves the tissues and enters the blood, both gases moving down their pressure gradients.

45. It is very thin, and has a large surface area.

46. There are very many capillaries, and the blood cells move through them in single file.

47. d.   **48.** d.   **49.** b.   **50.** c.

51. External respiration.

52., 53. and **54.**

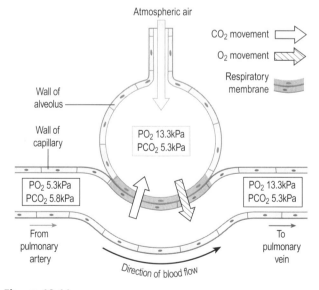

Figure 10.11

55. Internal respiration.

56. and **57.**

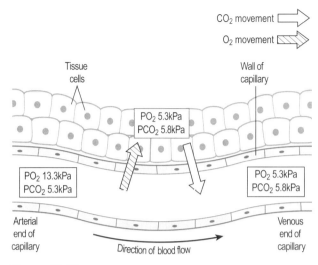

Figure 10.12

58.

Statement		Reason
A. Carbon dioxide diffuses from the body tissues into the bloodstream	because	PCO$_2$ is **higher** in the tissues than in the bloodstream.
B. Tissue levels of oxygen are lower than capillary blood levels	because	body cells are constantly **using oxygen**.
C. Oxygen diffuses out of the capillary into the tissue cells	because	**oxygen levels are higher in the capillary than in the tissues**.
D. Oxygen levels at the venous end of the capillary are lower than the arterial end	because	oxygen diffuses into the tissues as the blood flows through the capillary. **True**
E. Carbon dioxide moves out of the pulmonary capillaries into the alveoli	because	**carbon dioxide levels are higher in the capillaries than in the alveoli**.

59. a. CO_2, b. CO_2, c. O_2, d. CO_2, e. O_2, f. both, g. O_2, h. O_2, i. CO_2.

60.

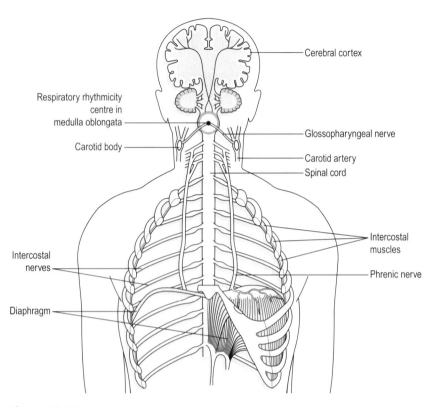

Figure 10.13

61. a. D; b. I; c. I; d. I; e. I; f. D; g. D; h. D; i. I; j. D.

62. d. **63.** a., c. **64.** d. **65.** a.

66.
  a. high levels of CO_2 in the blood
  b. low levels of O_2 in the tissues
  c. low levels of O_2 in the blood
  d. haemoglobin bound to carbon dioxide
  e. haemoglobin bound to oxygen.

Nutrition

ANSWERS

1. Carbohydrates, proteins, fats, vitamins, minerals and water.

2. and 3. a. 16.6, underweight; b. 27.8 overweight; c. 19.5 normal; d. 31.1 obese.

4.

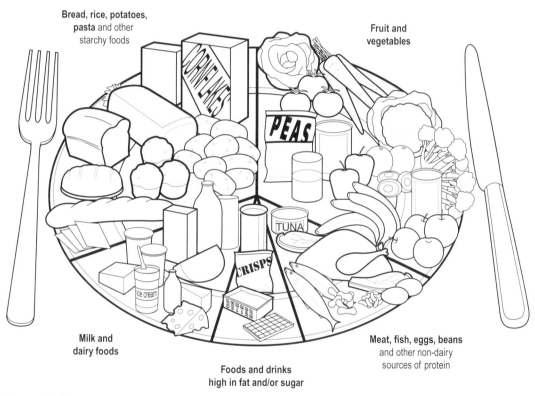

Figure 11.1

5. d.   6. d.   7. b., c., d.   8. a., c.

9. Contains all nutrients required for health in appropriate proportions.

10. Those that must be eaten in the diet because the body cannot sythesize them.

11. Those that the body can synthesize for itself.

12. Biological value is the extent to which food contains all essential amino acids in correct proportions.

13. Carbon, hydrogen, oxygen and nitrogen.

14. Iron, copper, zinc, iodine, sulphur and phosphate.

15. Growth and repair of body tissues; an alternative energy source when carbohydrates and fats are not available; building blocks for synthesis of enzymes, plasma proteins, antibodies.

16. The three elements that make up fat are **carbon, hydrogen** and **oxygen**. Fats are usually divided into two groups: **saturated** fats are found in foods from animal sources, such as **meat, fish** and **eggs**. The second group, the **unsaturated** fats, are found in vegetable oils. Fat (adipose) tissue is laid down under the skin, where it acts as an **insulator**. It is also found around the kidneys, where its function is to **support** these organs. Fat depots in the body are important as **energy** sources. Certain hormones, such as **steroids,** e.g. cortisone, are synthesized

283

from the fatty precursor **cholesterol**, also found in the cell membrane. In addition, certain substances are absorbed with fat in the intestine, a significant example being the **fat soluble vitamins**, which are essential for health despite being required only in tiny amounts. Fats in a meal have the direct effect of **slowing** gastric emptying and **delaying** the return of a feeling of hunger.

17. and **18. Table 11.1** Vitamin sources

Vitamin	Main sources
A (<u>fat soluble</u>)	Cream, egg yolk, liver, fish oil, milk, cheese, butter
B_1 (thiamine)	Nuts, egg yolk, yeast, liver, legumes, meat, cereal germ
B_2 (riboflavine)	Yeast, green vegetables, milk, liver, egg yolk, cheese
Folate (folic acid)	Liver, kidney, leafy green vegetables, yeast
Niacin	Liver, cheese, yeast, eggs, fish, nuts, whole cereal
B_6 (pyridoxine)	Egg yolk, peas, beans, yeast, meat, liver
B_{12} (cyanocobalamin)	Liver, meat, eggs, milk, fermented products
Pantothenic acid	Many foods
Biotin	Yeast, egg yolk, liver, kidney, tomatoes
C	Fresh citrus fruit
D (<u>fat soluble</u>)	Animal fats, e.g. eggs, butter, cheese, fish oils
E (<u>fat soluble</u>)	Nuts, egg yolk, wheat germ, whole cereal, milk, butter
K (<u>fat soluble</u>)	Fish, liver, leafy green vegetables and fruit

19. a. Vitamins C and E; b. vitamin C; c. vitamin A; d. vitamin B_6; e. vitamin A; f. vitamins B_1, B_2 and biotin; g. vitamin K; h. vitamins B_6, B_{12}, folate (folic acid); i. pantothenic acid, vitamin B_6; j. vitamin D; k. niacin; l. vitamin B_{12}.

20. Because vitamin A is a fat soluble vitamin, its absorption can be reduced if **bile** secretion into the gastrointestinal tract is lower than normal. The first sign of deficiency is **night blindness**, and this may be followed by **conjunctival ulceration**. On the other hand, the B complex vitamins are water soluble. Most of them are involved in **biochemical release of energy**. Thiamine deficiency is associated with **beriberi**, and niacin inadequacy leads to **pellagra**. Folic acid is required for **DNA** synthesis, and is therefore often prescribed as a supplement in pregnancy. Deficiency of vitamin B_{12} typically leads to **megaloblastic** anaemia, because it is needed for DNA synthesis, and is usually associated with lack of **intrinsic factor** in the gastrointestinal tract.

Vitamin C is needed for **connective tissue synthesis**. One of the first signs of deficiency of this vitamin is therefore loosening of the teeth, due to **defective gum tissue**. Vitamin C is destroyed by **heat**.

Lack of vitamin D causes **osteomalacia** in adults, and **rickets** in children. Vitamin E deficiency results in **haemolytic** anaemia, because the **cell membrane** of red blood cells is damaged.

Vitamin K deficiency leads to problems with **blood coagulation**.

21. Table 11.2 Functions of minerals

	Calcium	Phosphate	Sodium	Potassium	Iron	Iodine
Needed for haemoglobin synthesis					✓	
Used in thyroxine manufacture						✓
Most abundant cation outside cells			✓			
99% of body stock is found in bones	✓					
Most abundant cation inside cells				✓		
May be added to table salt						✓
Vitamin D is needed for use	✓	✓				
Involved in muscle contraction	✓		✓	✓		
Used to make high-energy ATP		✓				
Needed for normal blood clotting	✓					
Required for hardening of teeth	✓	✓				
Needed for normal nerve transmission			✓	✓		

22. c., d.   **23.** c., d.   **24.** a., b., c., d.   **25.** b.

26. a., b., c.   **27.** b., c.   **28.** b., c.   **29.** d.

30. a.   **31.** a., d.

32. Men – 60%; women – 55%.

33. a. F; b. T; c. T; d. F; e. T.

34. Non-starch polysaccharide.

35. Bulking diet and satisfying appetite; stimulating peristalsis of the intestines; attracting water that softens faeces; increases frequency of defaecation and prevents constipation; reduces incidence of certain gastrointestinal disorders.

36. Fruit, vegetables, whole cereals.

37. Constipation.

CHAPTER 1

The digestive system

1. and 2. The large intestine includes all regions of the colon and the rectum (see Figure 12.1).

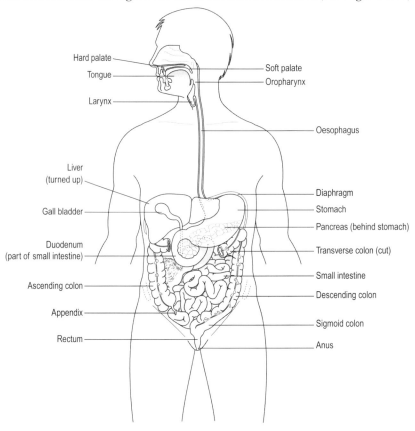

Hard palate
Tongue
Larynx
Soft palate
Oropharynx
Oesophagus
Liver
(turned up)
Gall bladder
Duodenum
(part of small intestine)
Ascending colon
Appendix
Rectum
Diaphragm
Stomach
Pancreas (behind stomach)
Transverse colon (cut)
Small intestine
Descending colon
Sigmoid colon
Anus

Figure 12.1

3. a. Ingestion – taking of substances into the alimentary canal by eating or drinking; b. propulsion – mixing and movement of substances along the alimentary tract; c. digestion – occurs through mechanical and chemical breakdown of food; d. absorption – the breakdown products of digestion are taken into the bloodstream for use by the body; e. elimination – excretion of substances that cannot be digested or absorbed as faeces.

4. Mechanical digestion is the physical squeezing, chopping or cutting of food in the gastrointestinal system, e.g. chewing by the teeth and churning in the stomach. Chemical digestion involves the breaking down of the molecules that make up the food into smaller ones that can be absorbed; this is accomplished by the gastrointestinal enzymes.

5. **Table 12.1** Organs of the alimentary tract and accessory organs

Organs of alimentary tract	Accessory organs
Mouth	Liver
Oesophagus	Gall bladder
Stomach	Pancreas
Small intestine	Sublingual glands
Large intestine	Submandibular glands
Rectum and anus	Parotid glands

6. and **7.**

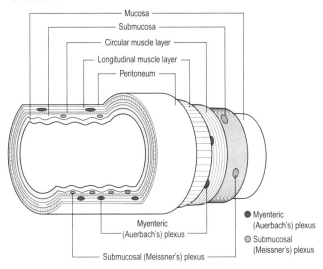

Figure 12.2

8. Sympathetic and parasympathetic.

9. b.   **10.** d.   **11.** b.   **12.** a.

13., 14. and **15.** This is columnar epithelium with goblet cells. Mucus lubricates the foodstuffs and protects the lining of the gastrointestinal tract.

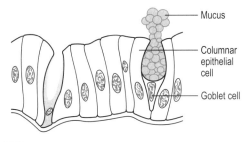

Figure 12.3

16. In areas where secretion and absorption occur, e.g. stomach and small/large intestine.

17.

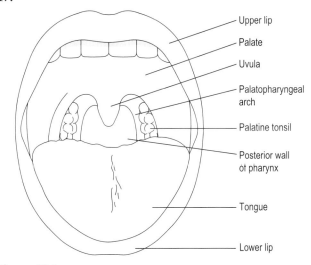

Figure 12.4

18. The hard palate forms the anterior part of the roof of the mouth while the soft palate lies posteriorly. The hard palate is formed by the palatine bones. The soft palate is muscular and curves downwards behind the hard palate blending with the walls of the pharynx at the sides.

19. and **20.** A = incisors, B = canines, C = premolars, D = molars; A and B: cutting and biting, C and D: grinding and chewing.

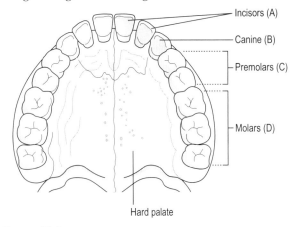

Figure 12.5

21. and **22.**

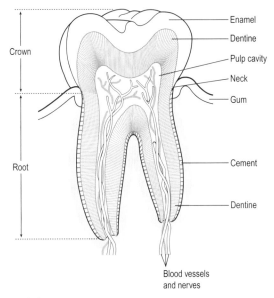

Figure 12.6

Cement – a bone-like substance that fixes the tooth in its socket.
Dentine – layer of the tooth lying below the enamel and surrounding the pulp cavity.
Enamel – very hard outer layer of the tooth that forms the crown.

23. Nerves, blood and lymph vessels.

24. Eight premolars (four at the bottom and four at the top), and four molars.

25. 6 years.   **26.** 21 years.

27.

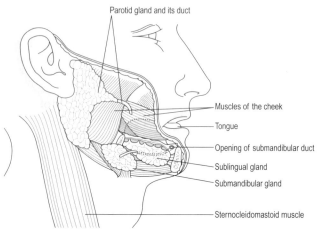

Figure 12.7

28. Parotid gland – duct opens into the mouth next to the second upper molar.
Submandibular gland – duct opens onto the floor of the mouth on each side of the frenulum.
Sublingual gland – numerous small ducts open into the floor of the mouth.

29.

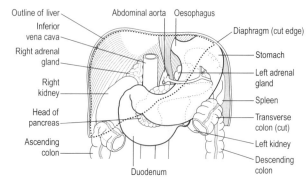

Figure 12.8

30.

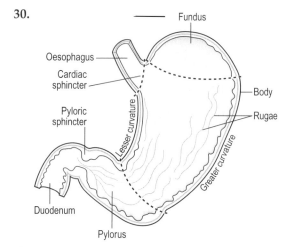

Figure 12.9

31. The stomach has three layers of smooth muscle. The inner layer consists of **longitudinal** fibres, the middle layer consists of **circular** fibres and the outer layer of **oblique** fibres.

32. This contributes to the efficient churning action of the stomach.

33. This is a thickened ring of circular smooth muscle which closes off the alimentary canal when it is constricted and opens as it relaxes allowing flow of contents along the alimentary tract. This valve-like action regulates the rate of flow along the tract optimizing the time for each stage of digestion to take place.

34. a. Hydrochloric acid, intrinsic factor; b. hydrochloric acid; c. mucus; d. hydrochloric acid; e. pepsinogens; f. pepsinogens; g. pepsinogens; h. intrinsic factor; i. hydrochloric acid; j. mucus; k. pepsinogens; l. hydrochloric acid.

35. c.   **36.** d.   **37.** b.   **38.** a.   **39.** a.

40. a. True. b. Chemical digestion in the stomach includes the action of **pepsin**, an enzyme that acts on proteins and breaks them down to smaller polypeptides. c. True. d. The stomach has **little** absorptive function; its environment is too acidic, and absorption cannot occur until the food has been neutralized in the intestines. e. True. f. Absorption of iron takes place **in the small intestine**; the acid environment of the stomach solubilizes iron salts, an essential step in iron absorption. g. True. h. The stomach regulates flow of liquidized food into the next part of the digestive tract, the duodenum, through the **pyloric** sphincter.

41. **Table 12.2** Characteristics of the duodenum, jejunum and ileum

	Duodenum	Jejunum	Ileum
Longest portion of the small intestine			✓
Curves around the head of the pancreas	✓		
Vitamin B$_{12}$ is absorbed here			✓
About 25 cm long	✓		
Middle section		✓	
Ends at the ileocaecal valve			✓
Flow in is regulated by the pyloric sphincter	✓		
Most digestion takes place here	✓		
About 2 m long		✓	
Flow from here enters the large intestine			✓
Bile passes into this section	✓		
The pancreas passes its secretions into this section	✓		
Villi present here	✓	✓	✓
Most absorption takes place here		✓	

42. Protein molecule that speeds up (catalyses) a specific biochemical reaction, recognized by the suffix ~ase (usually). The substrate binds to a specific receptor site allowing the reaction to proceed; once complete the product is released and the enzyme can carry out the same reaction again.

43. The smooth muscle of the alimentary tract wall is composed of two layers, an inner layer of longitudinal muscle and an outer one of circular muscle. Stimulation of the myenteric plexus results in contraction of both muscle layers followed by a period of relaxation, which moves the contents onwards along the alimentary tract in small waves.

44. and 45. (See Figure 12.10.) Mucosa forms the villi.

46. Monosaccharides and amino acids.

47. Digested fats (fatty acids and glycerol).

48. Aggregated lymphatic follicles (Peyer's patches).

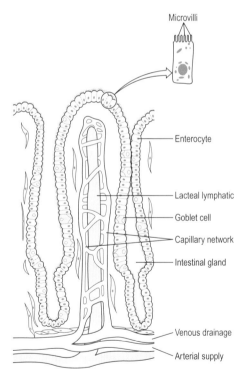

Figure 12.10

49. On a daily basis, the intestine secretes about **1500 ml** of intestinal juices, and its contents are **between 7.8 and 8.0**. In the small intestine, chemical digestion is completed and the end products are absorbed. The main enzyme secreted by the enterocytes is enterokinase, which **activates enzymes from the pancreas**. However, other enzymes from accessory structures are passed into the **duodenum** as well.

   The pancreas secretes **amylase**, which is important in reducing large sugar molecules to **disaccharides**. In addition, pancreatic lipase breaks down fats into **fatty acids and glycerol**, which can be absorbed in the intestine. The third major nutrient group, the proteins, are broken down to **dipeptides** by pancreatic **trypsin and chymotrypsin**. Pancreatic juice is also rich in **bicarbonate** ions, important in neutralizing the acid chyme from the stomach.

   Bile is made in the **liver**, stored in the **gall bladder**, and enters the intestine via the **hepatopancreatic sphincter**. It has a role to play in fat digestion by breaking fats into **tiny droplets**. This increases the action of lipases on the fat.

   Even after the multiple digestive actions of these enzymes, the digested proteins and carbohydrates are still not in a readily absorbable form, and digestion is completed by enzymes made by the **enterocytes**. Thus, the final stage of protein digestion produces **amino acids** and the final stage of carbohydrate digestion produces **monosaccharides**.

50. a., d.   51. b., c.   52. b.   53. c., d.   54. c.

55.

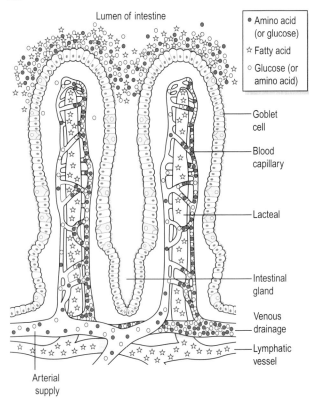

Figure 12.11

56. A, D, E and K in the lacteal (fat soluble); B and C in the blood.

57. Active transport.

58. Glucose, amino acids, fatty acids, glycerol, disaccharides, dipeptides, tripeptides.

59. Solitary lymph follicles and aggregated lymph follicles.

60. Parasympathetic (nervous system).

61. Parasympathetic (nervous system).

62. See Figure 12.12.

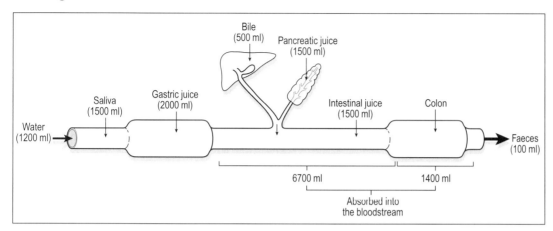

Figure 12.12

63. a., b. **64.** c. **65.** c. **66.** a., d. **67.** d.

68.

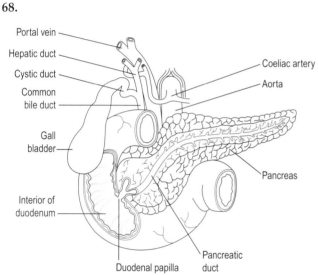

Figure 12.13

69. Bile and pancreatic juice.

70. Table 12.3 Functions of the pancreas

Exocrine functions	Endocrine functions
Secretion of enzymes	Secretion of insulin
Passes secretions into duodenum	Control of blood sugar levels
Secretions leave via the pancreatic duct	Secretion of hormones
Role is in digestion	Substances are passed directly into blood
Synthesis takes place in pancreatic alveoli	Secretion of glucagon
Secretions include amylase, lipase and proteases	Synthesis takes place in the pancreatic islets

71.

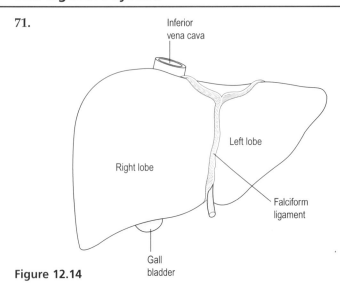

Figure 12.14

72. Diaphragm.

73.

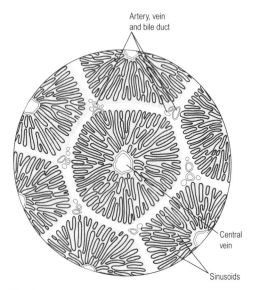

Figure 12.15

74. The arterial supply is from the hepatic artery and nourishes the liver tissues; the venous supply is the hepatic portal vein which is bringing blood from the intestines, to be purified before being returned to the venous circulation.

75. The liver is involved in the metabolism of carbohydrates; it converts glucose to **glycogen** for storage; the hormone that is important for this is **insulin**. In the opposite reaction, glucose is released to meet the body's energy needs and the important hormone for this is **glucagon**. This action of the liver maintains the blood sugar levels within close limits. Other metabolic processes include the formation of waste, including **urea** from the breakdown of protein, and **uric acid** from the breakdown of nucleic acids. Transamination is the process by which **new amino acids** are made from **carbohydrates**. Proteins are also made here; two important groups of proteins, found in the blood, are the **clotting proteins** and the **plasma proteins**.

The liver detoxifies many ingested chemicals, including **alcohol** and **drugs**. It also breaks down some of the body's own products, such as **hormones**. Red blood cells and other cellular material such as microbes are broken down in the **Kupffer cells**. It synthesizes vitamin **A** from **carotene**, a provitamin found in plants such as carrots, and stores it, along with other vitamins. The liver is also the main storage site of **iron** (essential for haemoglobin synthesis).

The liver makes **bile**, which is stored in the gall bladder and important in digestion of **fats**. Bile salts are important for **emulsifying fats** in the small intestine, and are themselves reabsorbed from the gut and returned to the liver in the **blood**. This is called the **enterohepatic** circulation, and helps to conserve the body's store of bile salts. Bilirubin is released when **red blood cells** are broken down (this occurs mainly in the **spleen** and the **liver**). Bilirubin is not very soluble, so to increase its water solubility so that it can be excreted in the bile, it is conjugated with **glucuronic acid**. On its passage through the intestine, it is converted by bacteria to **stercobilin**, which is excreted in the faeces; some is, however, reabsorbed and excreted in the urine as **urobilinogen**. If levels of bilirubin in the blood are high, its yellow colour is seen in the tissues as **jaundice**.

76., 77. and **78.**

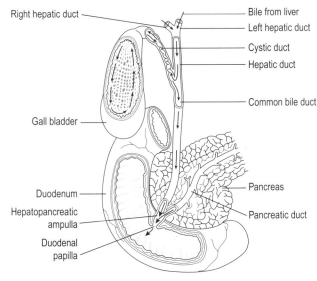

Figure 12.16

79. c.　**80.** a.　**81.** d.　**82.** c.　**83.** a.　**84.** b.

85. Catabolism is the breaking down of large molecules into smaller ones, often to release stored energy.

86. Anabolism is the synthesis of large molecules from smaller ones, usually requiring energy.

87. A kilocalorie is the amount of heat (energy) needed to raise the temperature of 1 litre of water by 1 degree Celsius. It is equivalent to 4.184 kilojoules.

88. d.　**89.** b.　**90.** d.　**91.** a.　**92.** d.　**93.** c.

94. and **95.**

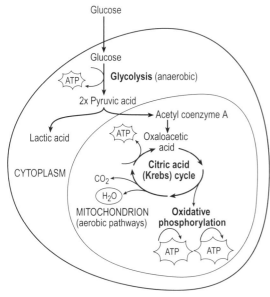

Figure 12.17

96. and **97.**

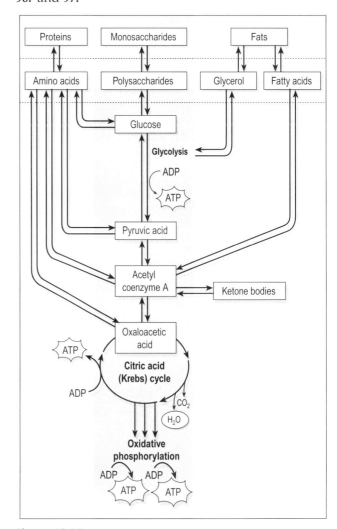

Figure 12.18

98. Oxygen.

The urinary system

1. and **2.**

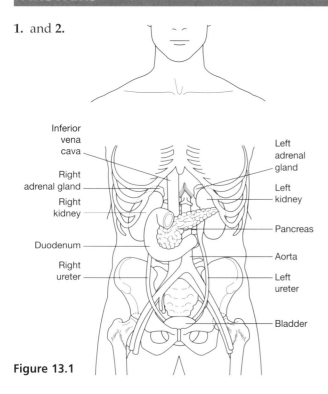

Figure 13.1

4. and **5.**

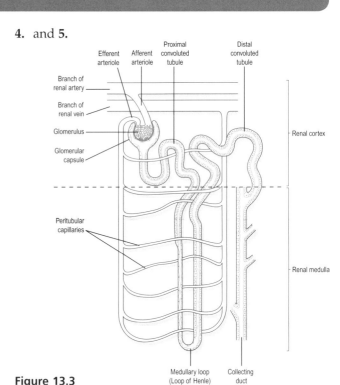

Figure 13.3

6. Collecting ducts, apex of renal papilla, minor calyx, major calyx, renal pelvis and ureter.

3.

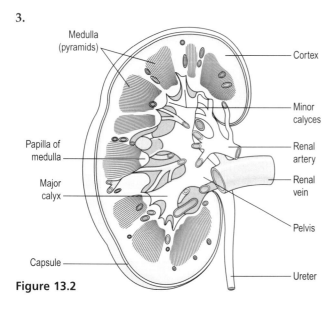

Figure 13.2

7.

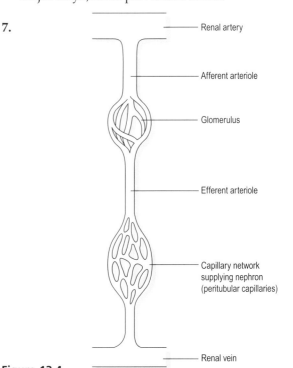

Figure 13.4

8. Mesangial cells.

9. d.   **10.** c.   **11.** a.   **12.** a.   **13.** d.   **14.** c.   **15.** a.

16. Filtration, reabsorption, secretion.

17. and **18**.

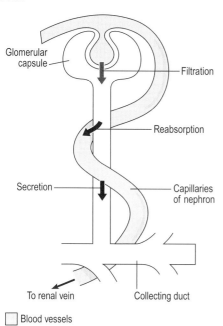

Figure 13.5

19. **Table 13.1** Characteristics of normal urine

Colour	Amber
Specific gravity	1020–1030
pH	6 (normal range = 4.5–8)
Average daily volume	1000–1500 ml

20. **Table 13.2** Normal constituents of glomerular filtrate and urine

Constituent of blood	Presence in glomerular filtrate	Presence in urine
Water	Normal	Normal
Sodium	Normal	Normal
Potassium	Normal	Normal
Glucose	Normal	Abnormal
Urea	Normal	Normal
Creatinine	Normal	Normal
Proteins	Abnormal	Abnormal
Uric acid	Normal	Normal
Red blood cells	Abnormal	Abnormal
White blood cells	Abnormal	Abnormal
Platelets	Abnormal	Abnormal

21. Water is excreted through the lungs in **saturated expired air**, through the skin as **sweat** and via the kidneys as the main constituent of **urine**. Of these three, the most important in controlling fluid balance are the **kidneys**. The minimum urinary output required to excrete the body's waste products is about **500 ml** per day. The volume in excess of this is controlled mainly by the hormone **ADH (antidiuretic hormone)**. Sensory nerve cells, called **osmoreceptors**, detect changes in the osmotic pressure of the blood. They are situated in the **hypothalamus**. When the osmotic pressure increases, secretion of ADH is **increased** and water is **reabsorbed** by the distal collecting tubules and collecting ducts. These actions result in the osmotic pressure of the blood being **decreased**. This control system maintains osmotic pressure of the blood within a narrow range and is known as a **negative feedback** system.

22. Sodium (g)
Potassium (e)
Water (h)
Increased (d)
Volume (i)
Vasoconstriction (c)
Renin (a)
ACE (angiotensin converting enzyme) (b)
Aldosterone (f).

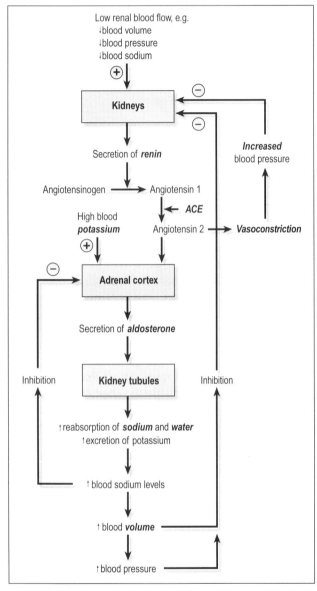

Figure 13.6

23. (j) Stimulates; (k) inhibits; (l) inhibits; (m) inhibits.

24. Aldosterone, atrial natriuretic peptide (ANP).

25. **Table 13.3** The sites of production of substances that influence the composition of urine

Substance	Site of production
Antidiuretic hormone	Posterior lobe of the pituitary gland
Aldosterone	Adrenal gland
Angiotensin converting enzyme	Lungs and proximal convoluted tubules
Renin	Afferent arteriole of the nephron
Angiotensinogen	Liver
Atrial natriuretic peptide	Atrial walls in the heart

26. At the upper end the ureter is continuous with the [renal] pelvis of the kidney. It passes downwards through the **abdominal** cavity, behind the **peritoneum** in front of the psoas muscle into the **pelvic** cavity and enters the **bladder**. Urine is propelled down the ureter by the process of **peristalsis**.

   Organs that lie behind the peritoneum can also be described as **retroperitoneal**.

27. a.   28. c.   29. a.   30. c.

31.

Inner layer (lined with transitional epithelium)

Ureter

Smooth muscle layer

Outer layer of bladder wall (fibrous tissue)

Figure 13.7

32. The ureters pass obliquely through the bladder wall so that when the bladder fills and the pressure inside rises the openings are compressed. This prevents backflow, or reflux, of urine (and potentially infection) into the ureters.

33. and **34.**

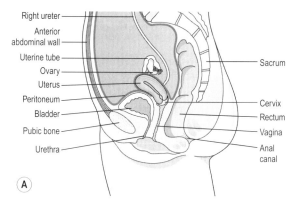

Ⓐ

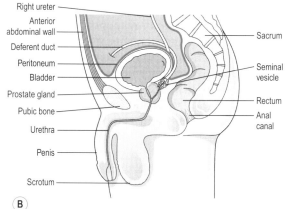

Ⓑ

Figure 13.8

35. The bladder acts as a **reservoir** for urine. When empty, its shape resembles a **pear** and it becomes more **oval** as it fills. The posterior surface is the **base** and the bladder opens into the urethra at its lowest point, the **neck**. The bladder wall is composed of three layers. The outer layer is composed of **connective tissue** and contains **blood** and **lymphatic** vessels. The muscular layer is formed by **smooth** muscle arranged in **three** layers. Collectively this is called the **detrusor** and when it contracts the bladder **empties**. The inner layer is the **mucosa** and it is lined with **transitional epithelium**. Three orifices on the posterior bladder wall form the **trigone**. The two upper openings are formed when each **ureter** enters the bladder and the lower one is the opening of the **urethra**.

36. d.   **37.** c.   **38.** b.   **39.** a.   **40.** c.   **41.** b.   **42.** c.

43. As the bladder fills and becomes distended, receptors in the wall are stimulated by **stretching**. In infants this initiates a **spinal reflex** and micturition occurs as nerve impulses to the bladder cause **contraction** of the **detrusor** muscle and **relaxation** of the **internal** urethral sphincter. When the nervous system is fully developed the micturition reflex is stimulated but sensory impulses pass upwards to the **brain**. By conscious effort, the reflex can be **over-ridden**. In addition to the processes involved in infants, there is **voluntary** relaxation of the **external** urethral sphincter.

44. Secretion of excessive volumes of urine.

45. Presence of sugar in urine.

46. Excessive thirst.

47. Presence of ketones in urine.

48. Blood in the urine.

49. Protein in the urine.

50. Plasma proteins; erythrocytes; leukocytes.

51. When glucose levels in the filtrate exceed the transport maximum of the kidneys no more glucose can be reabsorbed. It is therefore excreted in the urine together with large volumes of water. This leads to polyuria, polydipsia and dehydration.

52. It becomes excessive due to the absence of ADH.

53. They reduce blood pressure (antihypertensive agents).

54. a. Strong peristaltic activity of the ureteric smooth muscle occurs in an effort to dislodge the stone causing acute ischaemic pain. b. Renal colic. c. Accumulation of urine above the blockage distends the ureter and predisposes to infection, backflow to the renal pelvis leads to kidney damage. d. Calculi (singular = calculus). e. Lithotripsy.

55. The shorter female urethra is closer to the anus and the moist perineal conditions there.

ANSWERS

1., 2. and **3.**

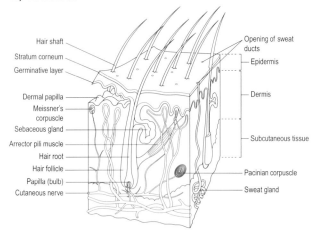

Hair shaft
Stratum corneum
Germinative layer
Dermal papilla
Meissner's corpuscle
Sebaceous gland
Arrector pili muscle
Hair root
Hair follicle
Papilla (bulb)
Cutaneous nerve

Opening of sweat ducts
Epidermis
Dermis
Subcutaneous tissue
Pacinian corpuscle
Sweat gland

Figure 14.1

4. c. **5.** b. **6.** a. **7.** a., b., c. **8.** c., d. **9.** c.

10. a. Sensory nerve endings; b. conduction; c. vasodilation; d. vitamin D; e. Langerhans cells; f. evaporation; g. absorption; h. convection; i. non-specific defence mechanism.

11. a. Gain; b. Loss; c. Loss; d. Loss; e. Loss; f. Gain; g. Loss; h. Loss.

12. a. Insensible water loss is around 500 ml per day; b. The distribution of sensory receptors in the skin is different, meaning that some areas are more sensitive than others; c. Correct; d. The skin is involved in the formation of vitamin D.

13. The temperature regulating centre is situated in the **hypothalamus** and is responsive to the temperature of circulating **blood**. When body temperature rises, sweat glands are stimulated by the **sympathetic nervous system**. The **vasomotor** centre in the medulla oblongata controls the diameter of small arteries and **arterioles** and therefore the amount of **blood** circulating in the dermis. When body temperature rises the skin capillaries **dilate** and extra blood near the surface

increases heat loss by **radiation**, **convection** and **conduction**. The skin is warm and **pink** in colour. When body temperature falls, arteriolar vasoconstriction conserves heat and the skin is **whiter / paler** and feels cool.

Fever is often the result of **infection**. During this process there is release of chemicals, also called **pyrogens**, from damaged tissue. These chemicals act on the **hypothalamus/temperature regulating centre** which releases prostaglandins that reset the temperature thermostat to a **higher** temperature. The body responds by activating heat promoting mechanisms, e.g. **shivering** and **vasoconstriction**, until the new temperature is reached. When the thermostat is reset to the normal level, heat loss mechanisms are activated. There is vasodilation and profuse **sweating** until body temperature returns to the normal range again.

14. Table 14.1 Factors affecting the rate of wound healing

	Promote wound healing	Impair wound healing
Systemic factors	Good nutritional status Good health	Infection Impaired immunity or illness
Local factors	Good blood supply	Contaminants

15. Healing that occurs when there is minimal loss of tissue and damaged skin edges are in close proximity.

16. Healing that occurs when there is destruction or loss of large amounts of tissue or when the edges of the wound cannot be brought together.

17. and **18.**

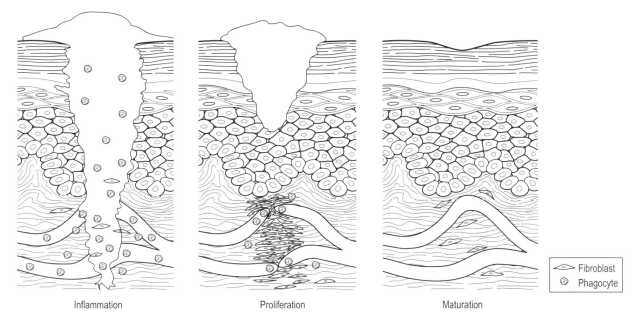

Inflammation

Proliferation

Maturation

Fibroblast
Phagocyte

Figure 14.2

19. a. Phagocytes   b. Fibroblasts   c. Slough
d. Phagocytes   e. Granulation tissue
f. Scar tissue.

20. b., c., d.   **21.** c., d.   **22.** b.   **23.** c.

24. Core body temperature less than 35 °C.

25. Vasconstriction occurs to conserve heat and blood supply to the skin is reduced and therefore there is less blood (containing the red pigment haemoglobin) there.

26. As body temperature drops, shivering occurs in an attempt to reverse heat loss. As body temperature decreases further heat loss continues and when heat conserving mechanisms fail, shivering stops.

ANSWERS

1. a. N/S, b. N/S, c. S, d. N/S, e. N/S, f. S, g. N/S, h. S, i. N/S; j. N/S; k. S, l. N/S.

2. c.   3. c.   4. a.   5. d.   6. c.

7. Phagocytosis.

8.

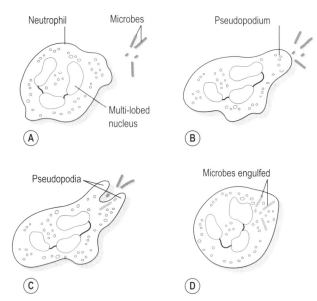

Figure 15.1

9. a. Neutrophil has been attracted towards invading bacteria; b. neutrophil is extending pseudopodium towards bacteria; c. the pseudopodia begin to encircle bacteria; d. bacteria are engulfed and will be destroyed.

10. and **11.** See Figure 15.2.
    A. Monocyte, a type of white blood cell, which migrates into the tissues in inflammation and differentiates into macrophages; B. macrophage, derived from a monocyte, a large efficient phagocyte that clears up tissue debris and microbes; C. mast cell, which makes and stores histamine in its cytoplasmic granules; D. neutrophil, a small phagocyte from the blood which is first on the scene in an acute inflammatory reaction.

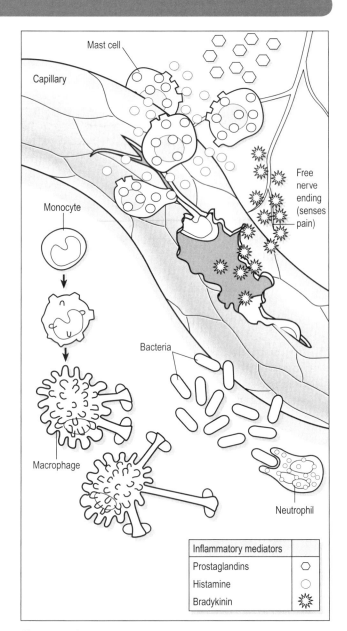

Figure 15.2

12. Redness, swelling, pain, heat, loss of function.

13.

A. Blood flow to an inflamed area increases	because	**Inflammatory mediators such as histamine** cause vasodilation.
B. An inflamed area swells	because	hydrostatic pressure at the arterial ends of local blood vessels **rises**.
C. White blood cells migrate into inflamed tissues	because	they are attracted by **inflammatory mediators such as prostaglandins and complement**.
D. The increased tissue temperature in inflammation is beneficial	because	it promotes the activity of phagocytes. **True**
E. The movement of fibrinogen from the bloodstream into inflamed and damaged tissues is beneficial	because	it **walls off the inflamed area and limits spread of infection.**

14. Table 15.1 Summary of some important inflammatory mediators

Substance	Made by	Trigger for release	Main actions
Histamine	Mast cells in tissues and basophils in blood; stored in cytoplasmic granules	Binding of antibody to mast cell or basophil plasma membrane	Vasodilation, itch, ↑ vascular permeability, degranulation, smooth muscle contraction (e.g. bronchoconstriction)
Serotonin	Platelets, mast cells and basophils; neurotransmitter in central nervous system	When platelets are activated and when mast cells/basophils degranulate	Vasoconstriction, ↑ vascular permeability
Prostaglandins	Synthesized as required from cell membranes	Many triggers, e.g. drugs, toxins, other mediators, trauma, hormones	Diverse, sometimes opposing, e.g. fever, pain, vasodilation or vasoconstriction, ↑ vascular permeability
Heparin	Liver, mast cells where it is stored in granules with histamine, basophils	When cells degranulate	Anticoagulant, maintaining blood supply to an inflamed area
Bradykinin	Tissues and blood	Blood clotting, trauma, inflammation	Pain by acting directly on free nerve endings, vasodilation

15. **Table 15.2** Features of the inflammatory response

Limits movement of the damaged area	f, g
Cushions the area of damaged tissue	g
Associated with chronic inflammation	e
Brings antibodies into inflamed tissues	d
Increases blood supply for tissue repair	h, i
Stimulates chemotaxis	c
Responsible for phagocytosis	a
Increases the temperature of damaged area	h, i
First protective cell to arrive at area of damaged tissue	a
Enhances phagocytosis of bacteria	a, c
Promotes activity of phagocytes	b, c

16. **Table 15.3** Lymphocyte characteristics

Characteristic	T-lymphocyte	B-lymphocyte
Shape of nucleus	Large, single	Large, single
Site of manufacture	Bone marrow	Bone marrow
Site of post-manufacture processing	Thymus gland	Bone marrow
Nature of immunity involved	Cell-mediated	Antibody-mediated
Specific or non-specific defence	Specific	Specific
Production of antibodies	No	Yes (as plasma cells)
Processing regulated by thymosin	Yes	No

17. **Table 15.4** The five types of antibody and their functions

Antibody type	Function
IgA	Present in body fluids (e.g. breast milk); prevents antigens crossing epithelial membranes and infecting deeper tissues
IgD	Made and displayed by B-lymphocytes to detect antigen
IgE	Made and displayed by mast cells and basophils and when it binds antigen, releases histamine from these cells
IgG	Largest commonest antibody; main antibody synthesized in secondary immune response
IgM	Potent activator of complement; found mainly in primary immune response

18. and 19. Macrophage is a non-specific phagocyte but in the immune response it presents antigenic fraction of antigen to unstimulated T-lymphocytes. Unspecialized T-lymphocytes have been processed to recognize only one antigen but have not yet encountered it; it will be activated by the macrophage showing it the antigen it is educated to look for.

Cytotoxic T-lymphocyte is one type of differentiated T-lymphocyte active in direct cell–cell killing, i.e. any cell that shows the target antigen will be destroyed.

Helper T-lymphocyte synthesizes chemicals to support other cells, and is also needed to interact with B-lymphocytes before B-lymphocytes can be activated to make antibody. Memory T-lymphocyte is long-lived and survives after infection is resolved; it will stimulate a faster and stronger response next time; this is the basis of immunity. Suppressor T-lymphocytes turn off the immune response once the threat has been dealt with.

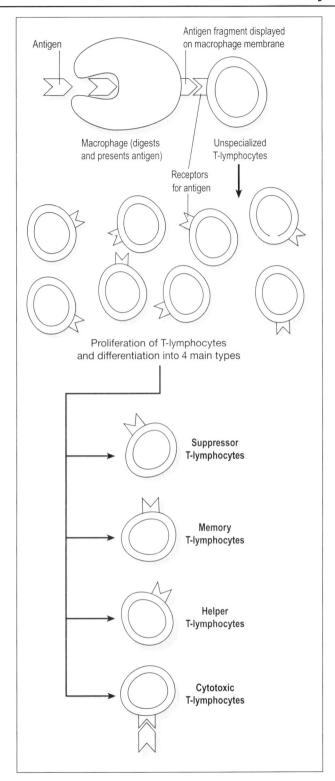

Figure 15.3

20. They are all the same, i.e. clonal expansion gives rise to different populations of T-cells that are all specific to the original antigen.

21. Clonal expansion.

22. and 23.

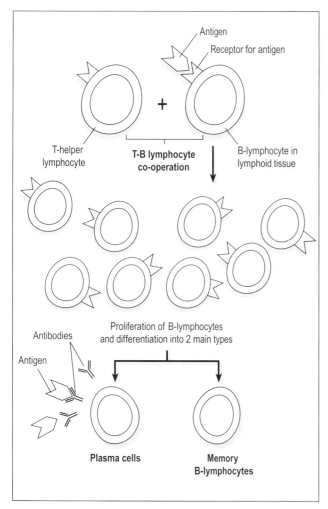

Figure 15.4

Helper T-lymphocyte is presenting antigen to the B-lymphocyte (i.e. is showing it the antigen it is looking for), and this stimulates the B-lymphocyte to divide and differentiate.

B-lymphocyte has been processed to recognize only one antigen, but has not yet encountered it; it will be activated by the helper T-cell showing it the antigen it is educated to look for.

Memory B-lymphocyte is long-lived and survives after infection is resolved; it will stimulate a faster and stronger response next time; this is the basis of immunity.

Plasma cell is derived from activated B-lymphocytes and makes antibody to the original antigen.

24. T–B-lymphocyte co-operation, essential for B-cell activation.

25. They are all the same; activated T lymphocytes only activate the B-cells carrying the same surface receptor as themselves, and so the B-cells make antibody only to the original antigen.

26. a.   27. d.   28. a., c.   29. b.   30. a.

31. When the body is exposed to an antigen for the first time, the immune response can be measured as antibody levels in the blood after about **2 weeks;** this is the **primary** response. Antibody levels fall thereafter, and do not rise again unless there is a second exposure to the same antigen, which stimulates a **secondary** response, which is different from the first in that it is much **faster** and antibody levels become much **higher**. After having been exposed to an antigen, an individual may develop immunity to it, provided he has produced a population of **memory** cells.

32. **Table 15.5** The four types of acquired immunity

Characteristic	Active natural	Active artificial	Passive natural	Passive artificial
An example is a baby's consumption of antibodies in its mother's milk			✓	
Long-lived protection	✓	✓		
Involves production of memory cells	✓	✓		
An example is vaccination		✓		
Short-lived protection			✓	✓
An example is infusion of antibodies				✓
Involves production of antibodies by the individual	✓	✓		
An example is a child catching chickenpox at school	✓			
Specific	✓	✓	✓	✓

1.

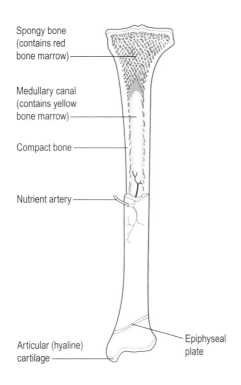

Spongy bone
(contains red
bone marrow)

Medullary canal
(contains yellow
bone marrow)

Compact bone

Nutrient artery

Articular (hyaline)
cartilage

Epiphyseal
plate

Figure 16.1

2. Periosteum.

3. a. Femur, tibia, fibula; b. carpals (wrist);
c. vertebrae; d. sternum, ribs; e. patella (knee cap).

4.

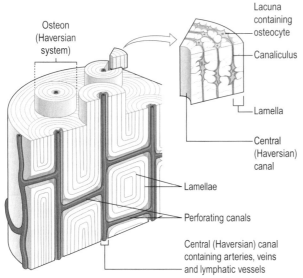

Osteon
(Haversian
system)

Lacuna
containing
osteocyte

Canaliculus

Lamella

Central
(Haversian)
canal

Lamellae

Perforating canals

Central (Haversian) canal
containing arteries, veins
and lymphatic vessels

5. **Table 16.1** Characteristics of bone

Spongy bone	Looks like a honeycomb to the naked eye
Osteoblasts	Cells that lay down bone
Osteoclasts	Cells that break down bone
Osteon	Haversian system
Spongy bone	Cancellous bone
Cortical bone	Compact bone
Trabeculae	Form the framework of spongy bone
Chrondroblasts	Cells that lay down cartilage
Osteocytes	Mature osteoblasts
Interstitial lamellae	Remains of old osteons
Lacunae	Spaces between lamellae that contain osteocytes
Red bone marrow	Found mainly in spaces within spongy bone
Flat bones	Develop from membrane models
Sesamoid bones	Develop from tendon models
Long bones	Develop from cartilage models

6. Bone tissue develops in the fetus from **connective** tissue models. This process is called **osteogenesis/ ossification** and is **incomplete** at birth. The main constituent of bone is **calcium salts**, and the organic component is primarily **collagen**. During life, bone growth is stimulated by **thyroxine/oestrogen**, but its density is decreased by **lack of exercise**.

7. and **8.** a. Fossa; b. border; c. trochanter; d. foramen; e. condyle; f. meatus; g. suture; h. fissure; i. crest/spine; j. facet; k. sinus.

S		C					T		T
I		O		C	R	E	S	T	R
N		N		C					O
U		D		A					C
S		Y	F	O	S	S	A		H
		L							A
B	M	E	A	T	U	S		E	N
O					R			S	T
R	F	I	S	S	U	R	E	P	E
D				T				I	R
E			U					N	
R		S		F	O	R	A	M	E	N

9. **Table 16.2** Types of fractures

Type of fracture	Characteristics
Simple	Bone ends do not protrude through the skin
Compound	Bone ends protrude through the skin
Pathological	Fracture of the bone weakened by disease

10. a. **11.** b. **12.** d. **13.** a.

14. Presence of tissue fragments between the bone ends; poor blood supply; poor alignment of bone ends; mobility of the bone ends.

15. and **16.** (See Figure 16.3.)

17. and **18.** (See Figure 16.4.)

19. a. Palatine; b. maxilla; c. vomer; d. occipital; e. temporal; f. sphenoid; g. ethmoid; h. temporal; i. occipital; j. lacrimal.

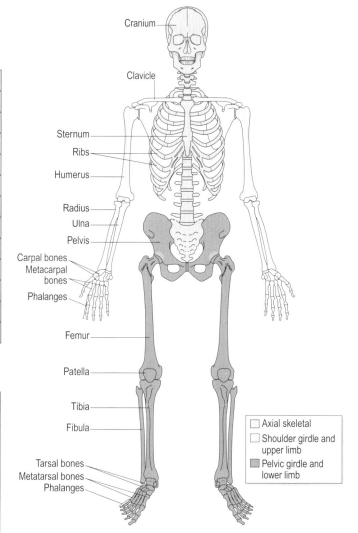

Cranium
Clavicle
Sternum
Ribs
Humerus
Radius
Ulna
Pelvis
Carpal bones
Metacarpal bones
Phalanges
Femur
Patella
Tibia
Fibula
Tarsal bones
Metatarsal bones
Phalanges

☐ Axial skeletal
☐ Shoulder girdle and upper limb
■ Pelvic girdle and lower limb

Figure 16.3

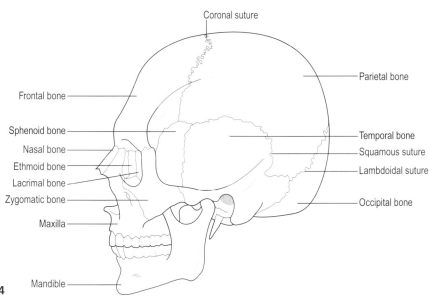

Coronal suture
Parietal bone
Frontal bone
Sphenoid bone
Nasal bone
Ethmoid bone
Lacrimal bone
Zygomatic bone
Maxilla
Temporal bone
Squamous suture
Lambdoidal suture
Occipital bone
Mandible

Figure 16.4

307

20. Sinuses contain **air** and are found in the **sphenoid, ethmoid, maxillary** and **frontal** bones. They all communicate with the **nasal cavity** and are lined with **ciliated epithelium**. Their functions are to give **resonance** to the voice and **lighten** the bones of the face and cranium.

Fontanelles are distinct **membranous** areas of the skull in infants and are present until **ossification** is complete and the skull bones fuse. The largest are the **anterior** fontanelle, present until **12–18** months, and the **posterior** fontanelle that usually closes over by **2–3** months of age. Their presence allows for moulding of the baby's **head** during childbirth.

21. and **23.**

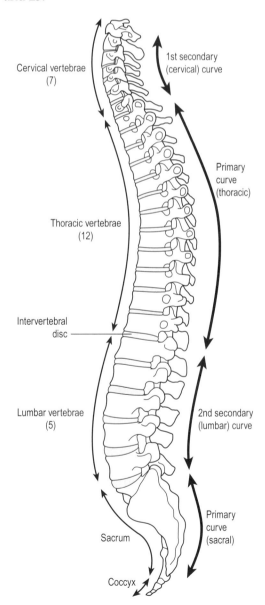

Figure 16.5

22. a. 7, b. 12, c. 5, d. 5, e. 4.

24.

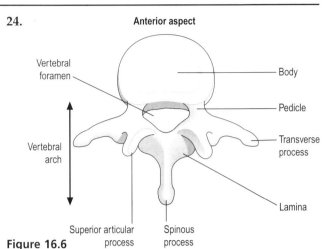

Figure 16.6

25. a. Coccyx; b. transverse foramen; c. thoracic vertebrae; d. intervertebral foramen; e. atlas; f. axis; g. sacrum; h. lumbar vertebrae; i. odontoid process; j. vertebral foramen; k. intervertebral disc; l. annulus fibrosus; m. nucleus pulposus.

26.

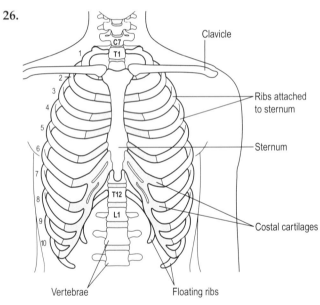

Figure 16.7

27. Intercostal nerves.

28.

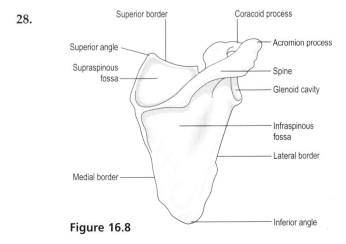

Figure 16.8

29.

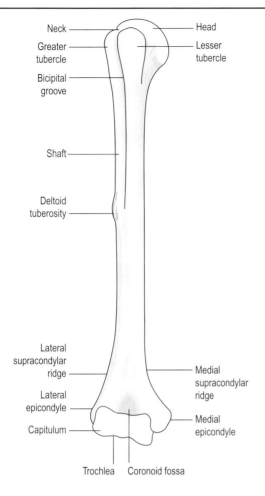

Figure 16.9

30. and 31.

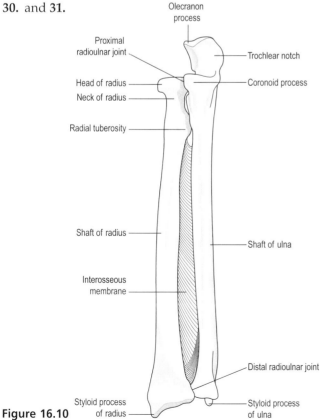

Figure 16.10

32. and 33.

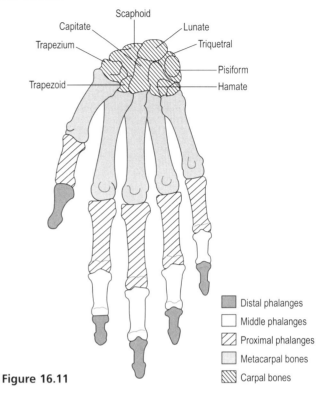

Figure 16.11

34. a. Humerus; b. scapula; c. ulna; d. scapula.

35. and 36.

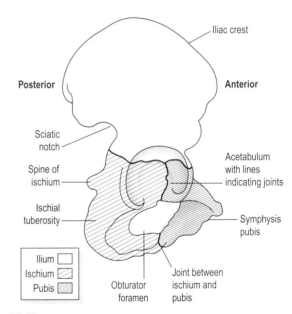

Figure 16.12

37. and 38.

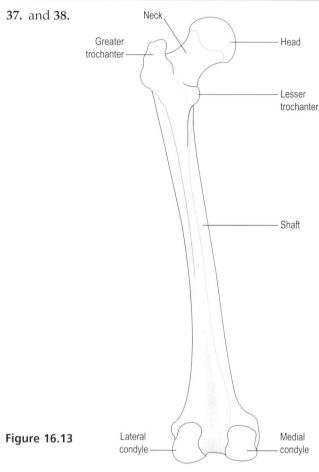

Figure 16.13

39. and 40.

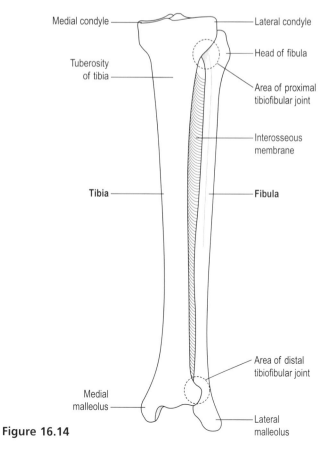

Figure 16.14

41. To stabilize and maintain the alignment of the tibia and fibula.

42. and 43.

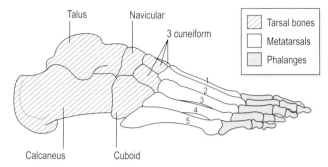

Figure 16.15

44. Flat foot is due to stretching and sagging of the ligaments and tendons that maintain the normal arched shape of the sole of the foot. It interferes with the normal 'springiness' of foot structure and the normal ability of the foot to distribute body weight while upright, and gives pain while standing, walking or running.

45. Table 16.3 Joints and movements

	Type (S, F or C)	Movement (I, SI, Fr)
Suture	F	I
Tooth in jaw	C	I
Shoulder joint	S	Fr
Symphysis pubis	C	SI
Knee joint	S	Fr
Interosseous membrane	F	SI
Hip joint	S	Fr
Joint between phalanges	S	Fr
Intervertebral discs	C	SI

46. a. Flexion; b. extension; c. abduction; d. adduction; e. circumduction; f. rotation; g. pronation; h. supination; i. inversion; j. eversion.

47.

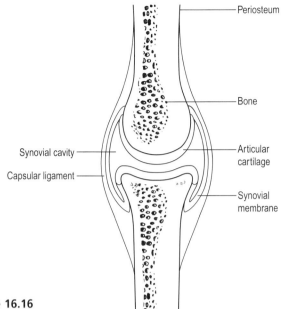

Figure 16.16

48. d.   **49.** a.   **50.** c.   **51.** a., c.

52. and **53.**

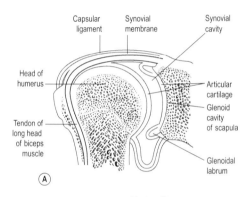

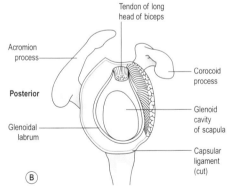

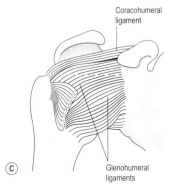

Figure 16.17

54. Ball and socket.

55. Generally, the more mobile the joint, the less stable it is.

56. Table 16.4 Muscles involved in movement at the shoulder joint

Movement	Muscle(s) involved
Flexion	Coracobrachialis, anterior fibres of deltoid and pectoralis major
Extension	Teres major, latissimus dorsi and posterior deltoid
Abduction	Deltoid
Adduction	Combined action of flexors and extensors
Circumduction	Flexors, extensors, abductors and adductors
Medial rotation	Pectoralis major, latissimus dorsi, teres major and anterior deltoid
Lateral rotation	Posterior fibres of deltoid

57. Hinge.

58. Table 16.5 Muscles involved in movement of the elbow

Movement	Muscle(s) involved
Flexion	Biceps and brachialis
Extension	Triceps

59. Hinge.

60. In the toes.

61. Table 16.6 Muscles involved in movement of the proximal and distal radioulnar joints and wrist

Movement of radioulnar joints	Muscle(s) involved
Pronation	Pronator teres
Supination	Supinator and biceps
Movement of the wrist	
Flexion	Flexor carpi radialis and flexor carpi ulnaris
Extension	Extensor carpi radialis (longis and brevis) and extensor carpi ulnaris
Abduction	Flexor and extensor carpi radialis
Adduction	Flexor and extensor carpi ulnaris

65. Table 16.7 Muscles involved in movement of the hip

Movement	Muscle(s) involved
Flexion	Psoas, iliacus, rectus femoris and sartorius
Extension	Gluteus maximus and hamstrings
Abduction	Gluteus medius and minimus, and sartorius
Adduction	Adductor group
Medial rotation	Adductor group, and gluteus medius and minimus
Lateral rotation	Quadriceps femoris, gluteus maximus, sartorius and, sometimes, adductor group

62. Ball and socket.

63. and **64.**

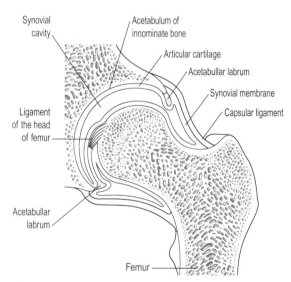

Figure 16.18

67. and 67.

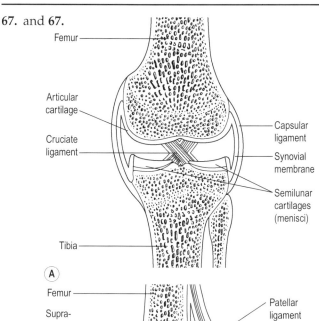

(A)

(B)

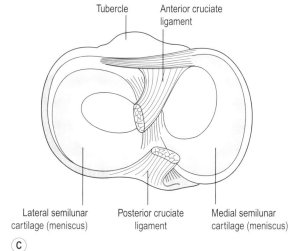

(C)

Figure 16.19

68. Hinge.

69. Table 16.8 Muscles involved in movement of the knee

Movement	Muscle(s) involved
Flexion	Gastrocnemius and hamstrings
Extension	Quadriceps femoris

70. 1. c.; 2. g.; 3. f.; 4. a.

71. Skeletal, cardiac and smooth muscle.

72.

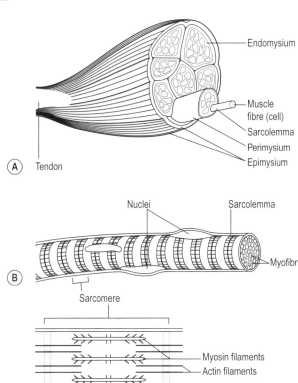

(A)

(B)

(C)

Figure 16.20

73. The functional unit of a skeletal muscle cell is the **sarcomere**. At each end of this unit are lines called **Z**-lines. Within the unit are two types of filament, thick filaments (made of **myosin**), and thin ones, made of **actin**. When the muscle cell is relaxed, these two filaments are not connected to each other. Contraction is initiated when an electrical impulse, called an **action potential**, passes along the cell membrane (also called the **sarcolemma**) of the muscle cell and penetrates deep into the sarcoplasm via the network of **channels** that run through the cell. This electrical stimulation causes **calcium** ions to be released from the **calcium stores** within the cell; these ions cause links, called **cross bridges**, to form between the thick and thin filaments. The filaments pull on each other, which causes the functional unit to **shorten** in length, pulling the **Z-lines** at either end towards one another. If enough units are stimulated to contract at the same time, the entire **muscle** will also **shorten (contract)**.

74. and **75.**

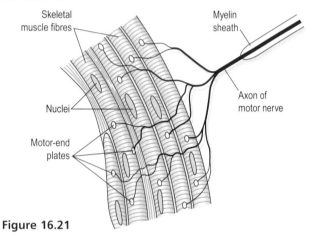

Figure 16.21

76. Acetylcholine.

77. A motor nerve and all the muscle fibres it supplies.

78. Muscle work with shortening of muscle.

79. Muscle work with no shortening of muscle but increased tension developed, as when trying to lift an immovable load.

80. The (usually proximal) attachment point of a muscle, which usually remains steady when the muscle contracts.

81. Increase in size of muscle, due to enlargement of individual muscle fibres.

82. Two muscles that work in opposition to each other across one or more joints.

83.

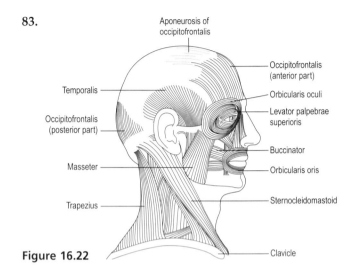

Figure 16.22

84. Table 16.9 Functions of muscles of the face and neck

Muscle	Paired/unpaired	Function
Occipitofrontalis	Unpaired	Raises the eyebrows
Levator palpebrae superioris	Paired	Raise the eyelids
Orbicularis oculi	Paired	Close the eyes and screw them up
Buccinator	Paired	Draw in the cheeks and expel air forcibly
Orbicularis oris	Unpaired	Closes the lips and involved in whistling
Masseter	Paired	Draw the mandible up to the maxilla for chewing
Temporalis	Paired	Close the mouth and involved in chewing
Pterygoid	Paired	Close the mouth and pull the lower jaw forwards
Sternocleidomastoid	Paired	Contraction of one: draws the head towards the shoulder
		Contraction of both: flexion of the cervical vertebrae drawing the sternum and clavicles upwards
Trapezius	Paired	Pull the head backwards, square the shoulders and control movements of the scapula when the shoulder is in use

85.

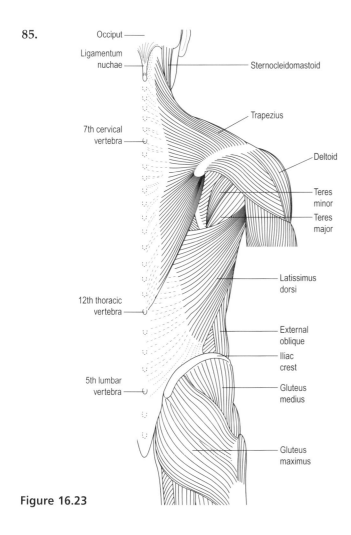

Figure 16.23

86. and 87.

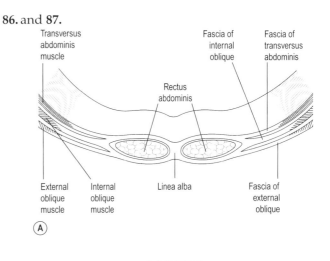

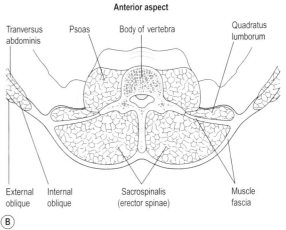

Figure 16.24

88.

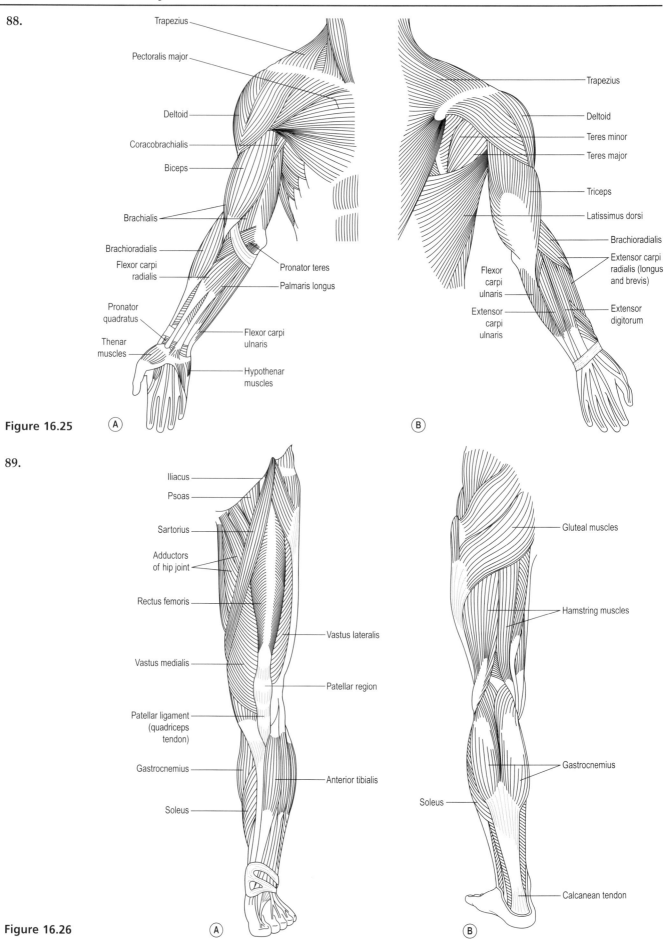

Figure 16.25

(A) (B)

89.

Figure 16.26

(A) (B)

90.

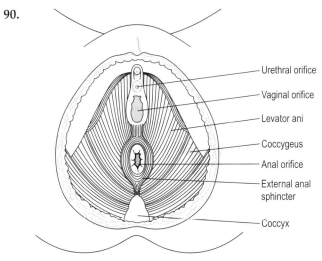

- Urethral orifice
- Vaginal orifice
- Levator ani
- Coccygeus
- Anal orifice
- External anal sphincter
- Coccyx

Figure 16.27

91. It acts as a support for the organs of the pelvis and maintains continence by resisting raised intrapelvic pressure during micturition and defaecation.

ANSWERS

1. a. Deoxyribonucleic acid.   b. Ribonucleic acid.

2. a.   **3.** c.   **4.** b.   **5.** b.

6. The nucleus contains the body's **genetic** material, in the form of DNA, which is built from nucleotides, each made up of three components: a **phosphate** group, the sugar **deoxyribose** and one of four **bases**. DNA is a double strand of nucleotides that resembles a **helix**, or twisted ladder. DNA and associated proteins, also called **histones**, are coiled together, forming **chromatin**. During cell division, the DNA becomes very tightly coiled and can be seen as **chromosomes** under the microscope. There are **23** pairs of them in most human cells. Each consists of many functional subunits called **genes**. Any given type of cell uses only part of the whole genetic code, also called the **genome**, to carry out its specific activities. Each **gene** contains the genetic code, or instructions, for the synthesis of one **protein**, that could, for example, be an **enzyme** needed to catalyse a particular chemical **reaction**, a hormone, or it may form part of the structure of a cell. The coded instructions have to be transferred to the **cytoplasm** of the cell, since that is where the organelles that make protein, the **ribosomes**, are found. DNA itself does not transfer, but a copy of the genetic code is made in the form of **mRNA**, which leaves the **nucleus**. When its instructions have been read and the new protein synthesized, the copy is destroyed.

7. and **8.**

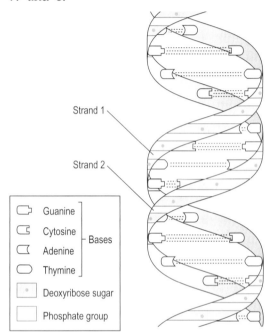

Strand 1

Strand 2

Guanine	
Cytosine	Bases
Adenine	
Thymine	
Deoxyribose sugar	
Phosphate group	

Figure 17.1

9. Table 17.1 The DNA code

DNA Strand 1	C	C	G	T	A	A	C	T	C	A	A	T	G	T
DNA Strand 2	G	G	C	A	T	T	G	A	G	T	T	A	C	A
mRNA	G	G	C	A	U	U	G	A	G	U	U	A	C	A

10. and **11.**

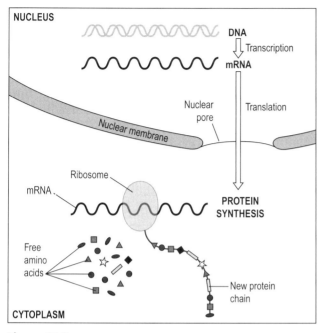

Figure 17.2

12. a. RNA; b. DNA; c. both; d. RNA; e. DNA;
f. RNA; g. RNA; h. RNA; i. both; j. DNA.

13. a. Translation takes place on the ribosomes in the
cytoplasm
b. True
c. A codon is a piece of RNA carrying information
d. True
e. Some new proteins are made for export,
e.g. insulin
f. Red blood cells have no nucleus, and gametes
carry only half
g. True
h. True.

14. Table 17.2 Characteristics of mitosis and meiosis

	Mitosis	**Meiosis**
One division or two?	One	Two
Daughter cells diploid or haploid?	Diploid	Haploid
Does crossing over take place?	No	Yes
Are daughter cells identical to parent?	Yes	No
Two or four cells produced?	Two	Four
Are daughter cells identical to one another?	Yes	No
Which process produces gametes?	–	Yes
Which process replaces damaged cells?	Yes	–

15. One chromosome of each pair is inherited from the mother and one from the father, so there are **two** copies of each gene in the cell. Two chromosomes of the same pair are called **homologues**, and the genes are present in paired sites called **alleles**.

    When the paired genes are identical, they are called **homozygous**, but if they are different forms they are called **heterozygous**. Dominant genes are always **expressed over recessive genes**. Individuals homozygous for a dominant gene **cannot** pass the recessive form on to their children, and individuals heterozygous for a gene **can** pass on either form of the gene to theirs.

16. Expression of genes in an individual, e.g. blue eyed or brown haired.

17. The genes present on an individual's chromosomes. Dominant genes are usually represented by a capital letter, recessive with the corresponding lower case letter.

18.
Box 17.1

	T	t
T	TT	Tt
t	Tt	tt

19. TT and Tt.

20. TT and tt.

21.
Box 17.2

	B	B
b	Bb	Bb
b	Bb	Bb

22. None (no bb).

23.
Box 17.3

	X^B	Y
X^b	$X^B X^b$	$X^b Y$
X^b	$X^B X^b$	$X^b Y$

24. 50:50.

25. 100% (both of them).

26. Carriers.

The reproductive system

ANSWERS

1. and 2.

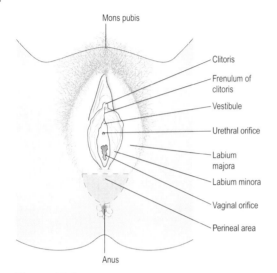

Figure 18.1

3. An incomplete fold of membrane partially occluding the vaginal opening.

4. Secretion of mucus to moisten the vulva.

5. and 6.

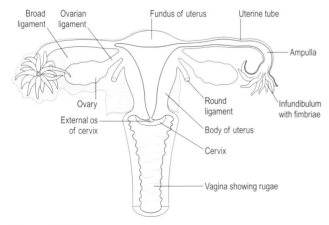

Figure 18.2

7. and 8.

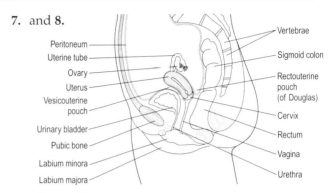

Figure 18.3

9. In the vagina, where the acid environment protects against ascending infection.

10. and 11.

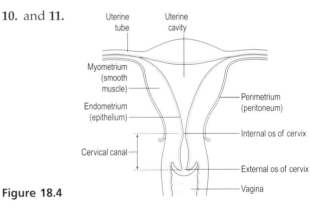

Figure 18.4

12. b.   **13.** a., d.   **14.** d.   **15.** d.

16., 17., 18. and 19.

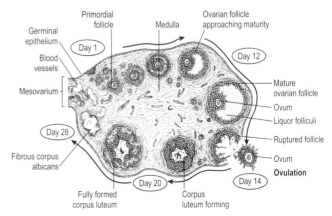

Figure 18.5

20. Maturation of the follicle is stimulated by **follicle stimulating hormone** released by the anterior pituitary, and oestrogen from the **follicular cells**. Ovulation is triggered by a surge of **luteinizing hormone**, which is secreted by the anterior pituitary. This release takes place a few **hours** before ovulation. After ovulation, the now empty follicle develops into the **corpus luteum**, and its main function is to secrete **progesterone and oestrogen**, which maintains the uterine lining in case fertilization and implantation occur. If pregnancy does occur, the embedded ovum supports the corpus luteum by producing **human chorionic gonadotrophin**, which keeps it functioning for the next 3 months or so, until the **placenta** is developed enough to take on this role. If pregnancy does not occur, the corpus luteum degenerates and forms a scar on the surface of the ovary called the **corpus albicans**.

21. Maturation of the uterus, uterine tubes and ovaries; beginning of the menstrual cycle; development of the breasts; growth of the pubic and axillary hair; increase in body height and pelvic width; deposition of subcutaneous fat, especially at hips and breasts.

22.

```
        ┌──────────────┐
        │ Hypothalamus │
        └──────┬───────┘
               ↓
   Luteinizing hormone
   releasing hormone (LHRH)
               ↓
        ┌───────────────────┐
        │ Anterior pituitary │
        └─────────┬─────────┘
           ┌──────┴──────┐
           ↓             ↓
  Follicle stimulating   Luteinizing
  hormone (FSH)          hormone (LH)
           ↓             ↓
  ┌────────────────┐  ┌──────────────┐
  │ Ovarian follicle│  │ Corpus luteum│
  └───────┬────────┘  └──────┬───────┘
          ↓                  ↓
      Oestrogen          Progesterone
```

23. a. LHRH from the hypothalamus causes anterior pituitary secretion of FSH and LH.
    b. FSH promotes ovarian follicular development and follicular secretion of oestrogen in the first half of the menstrual cycle.
    c. LH triggers ovulation, and stimulates development of the corpus luteum, which synthesizes progesterone.
    d. Oestrogen stimulates the development of the female secondary sexual characteristics at puberty, stimulates breast growth in pregnancy, stimulates proliferation of the endometrium in the first half of the menstrual cycle, and with progesterone, turns off LH and FSH production in the second half of the menstrual cycle (so that another ovum is not released until after menstruation).
    e. Progesterone with oestrogen promotes sexual changes at puberty; with oestrogen it regulates FSH and LH levels in the second half of the menstrual cycle; it maintains the thick vascular lining of the uterus in the second half of the menstrual cycle and during pregnancy and prevents menstruation.

24. and 25. See Figure 18.7.

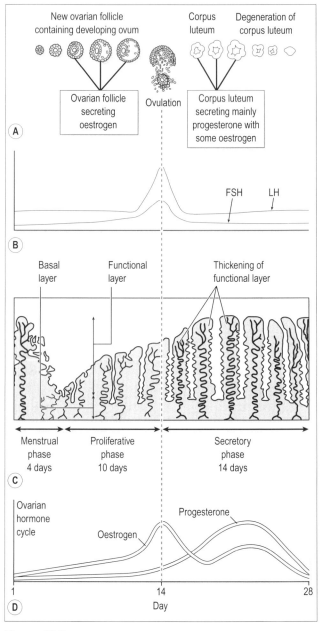

Figure 18.7

26. Anterior pituitary.

27. Ovulation (event E) is caused by the sudden surge in the levels of LH.

28. See the answer version of Figure 18.7.

29. Progesterone.

30. See the answer version of Figure 18.7.

31. They are being synthesized by the corpus luteum, which will begin to degenerate in the absence of pregnancy, and therefore oestrogen and progesterone levels will start to decline.

32. If pregnancy occurs, the developing embryo secretes human chorionic gonadotrophin (hCG), which maintains the corpus luteum for 3–4 months; during this time it continues to secrete the oestrogen and progesterone (on which the pregnancy depends). At the end of this time, the placenta is mature enough to maintain oestrogen and progesterone levels.

33.

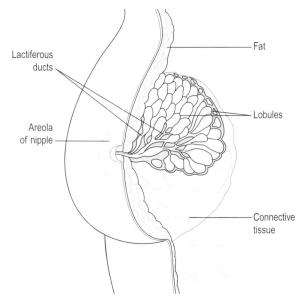

Figure 18.8

34. **Table 18.1** The effect of hormones on the breast

Statement	Hormone(s)
Stimulates body growth and development in puberty	Oestrogen and progesterone
Initiates release of milk	Oxytocin
Stimulates production of milk	Prolactin
Stimulates growth and development in pregnancy	Oestrogen and progesterone

35. and **36.**

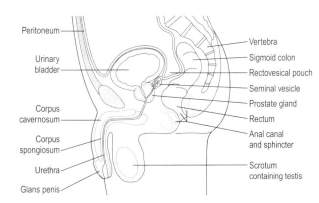

Figure 18.9

37., 38. and **39.** See Figure 18.10.

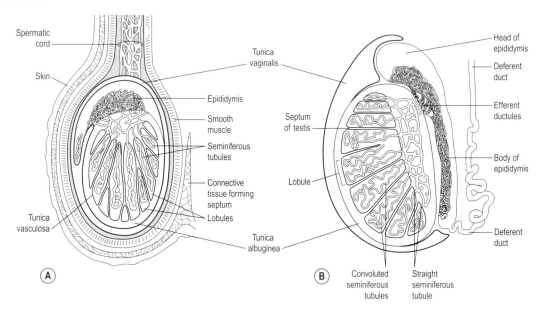

Figure 18.10

40. b.　**41.** a.　**42.** c.　**43.** a.

44. a., c.

45.

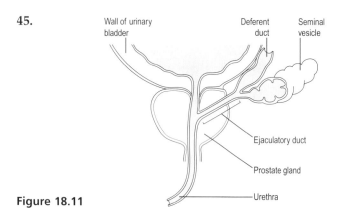

Figure 18.11

46. Table 18.2 Nature and function of semen

	% semen volume	Function
Spermatozoa	10	Sex cells to fertilize the ovum
Prostate secretions	30	Clotting enzyme to thicken sperm and increase the likelihood of sperm pooling at the cervix
Seminal vesicle secretions	60	Contain nutrients such as glucose to nourish sperm on their journey into the vagina; slightly alkaline to protect sperm from acid secretions of the vagina